STUDENT WORKBOOK
VOLUME 4

P H Y S I C S

FOR SCIENTISTS AND ENGINEERS SECOND EDITION

A STRATEGIC APPROACH

Randall D. Knight
California Polytechnic State University
San Luis Obispo

PEARSON

Addison
Wesley

San Francisco   Boston   New York
Capetown   Hong Kong   London   Madrid   Mexico City
Montreal   Munich   Paris   Singapore   Sydney   Tokyo   Toronto

Publisher: Adam Black, Ph.D.
Development Manager: Michael Gillespie
Development Editor: Alice Houston, Ph.D.
Project Editor: Martha Steele
Assistant Editor: Grace Joo
Media Producer: Deb Greco
Sr. Administrative Assistant: Cathy Glenn
Director of Marketing: Christy Lawrence
Executive Marketing Manager: Scott Dustan
Sr. Market Development Manager: Josh Frost
Market Development Associate: Jessica Lyons
Managing Editor: Corinne Benson
Production Supervisor: Nancy Tabor
Production Service: WestWords PMG
Illustrations: Precision Graphics
Text Design: Seventeenth Street Studios and WestWords PMG
Cover Design: Yvo Riezebos Design and Seventeenth Street Studios
Manufacturing Manager: Pam Augspurger
Text and Cover Printer: Edwards Brothers
Cover Image: Composite illustration by Yvo Riezebos Design; photo of spring by Bill Frymire/Masterfile

ISBN-13: 978-0-321-51629-9
ISBN-10: 0-321-51629-X

Table of Contents

Part VI   Electricity and Magnetism

Chapter 26   Electric Charges and Forces 26-1
Chapter 27   The Electric Field 27-1
Chapter 28   Gauss's Law 28-1
Chapter 29   The Electric Potential 29-1
Chapter 30   Potential and Field 30-1
Chapter 31   Current and Resistance 31-1
Chapter 32   Fundamentals of Circuits 32-1
Chapter 33   The Magnetic Field 33-1
Chapter 34   Electromagnetic Induction 34-1
Chapter 35   Electromagnetic Fields and Waves 35-1
Chapter 36   AC Circuits 36-1

Part VII   Relativity and Quantum Physics

Chapter 37   Relativity 37-1

Preface

Learning physics, just as learning any skill, requires regular practice of the basic techniques. That is what this *Student Workbook* is all about. The workbook consists of exercises that give you an opportunity to practice the ideas and techniques presented in the textbook and in class. These exercises are intended to be done on a daily basis, right after the topics have been discussed in class and are still fresh in your mind.

You will find that the exercises are nearly all *qualitative* rather than *quantitative*. They ask you to draw pictures, interpret graphs, use ratios, write short explanations, or provide other answers that do not involve significant calculations. The purpose of these exercises is to help you develop the basic thinking tools you'll later need for quantitative problem solving. Successful completion of the workbook exercises will prepare you to tackle the more quantitative end-of-chapter homework problems in the textbook. It is highly recommended that you do the workbook exercises *before* starting the end-of-chapter problems.

You will find that the exercises in this workbook are keyed to specific sections of the textbook in order to let you practice the new ideas introduced in that section. You should keep the text beside you as you work and refer to it often. You will usually find Tactics Boxes, figures, or examples in the textbook that are directly relevant to the exercises. When asked to draw figures or diagrams, you should attempt to draw them so that they look much like the figures and diagrams in the textbook.

Because the exercises go with specific sections of the text, you should answer them on the basis of information presented in *just* that section (and prior sections). You may have learned new ideas in Section 7 of a chapter, but you should not use those ideas when answering questions from Section 4. There will be ample opportunity in the Section 7 exercises to use that information there.

You will need a few "tools" to complete the exercises. Many of the exercises will ask you to *color code* your answers by drawing some items in black, others in red, and yet others in blue. You need to purchase a few colored pencils to do this. The author highly recommends that you work in pencil, rather than ink, so that you can easily erase. Few people produce work so free from errors that they can work in ink! In addition, you'll find that a small, easily carried six-inch ruler will come in handy for drawings and graphs.

As you work your way through the textbook and this workbook, you will find that physics is a way of *thinking* about how the world works and why things happen as they do. We will be interested primarily in finding relationships and seeking explanations, only secondarily in computing numerical answers. In many ways, the thinking tools developed in this workbook are what the course is all about. If you take the time to do these exercises regularly and to review the answers, in whatever form your instructor provides them, you will be well on your way to success in physics.

To the instructor: The exercises in this workbook can be used in many ways. You can have students work on some exercises in class as part of an active-learning strategy. Or you can do the same in recitation sections or laboratories. This approach allows you to discuss the answers immediately, to answer student questions, and to improvise follow-up exercises when needed. Having the students work in small groups (2 to 4 students) is highly recommended.

Alternatively, the exercises can be assigned as homework. The pages are perforated for easy tear-out, and the page breaks are in logical places so that you can assign the sections of a chapter that you would likely cover in one day of class. Exercises should be assigned immediately after presenting the relevant information in class and should be due at the beginning of the next class. Collecting them at the beginning of class, then going over two or three that are likely to cause difficulty, is an effective means of quickly reviewing major concepts from the previous class and launching a new discussion.

If the exercisees are used as homework, it is *essential* for students to receive *prompt* feedback. Ideally this would occur by having the exercises graded, with written comments, and returned at the next class meeting. Posting the answers on a course website also works. Lack of prompt feedback can negate much of the value of these exercises. Placing similar qualitative/ graphical questions on quizzes and exams, and telling students at the beginning of the term that you will do so, encourages students to take the exercises seriously and to check the answers.

The author has been successful with assigning *all* exercises in the workbook as homework, collecting and grading them every day through Chapter 4, then collecting and grading them on about one-third of subsequent days on a random basis. Student feedback from end-of-term questionnaires reveals three prevalent attitudes toward the workbook exercises:

i.   They think it is an unreasonable amount of work.

ii. They agree that the assignments force them to keep up and not get behind.

iii. They recognize, by the end of the term, that the workbook is a valuable learning tool.

However you choose to use these exercises, they will significantly strengthen your students' conceptual understanding of physics.

Following the workbook exercises are optional Dynamics Worksheets, Momentum Worksheets, and Energy Worksheets for use with end-of-chapter problems in Parts I and II of the textbook. Their use is recommended to help students acquire good problem-solving habits early in the course. End-of-chapter problems marked with the ✎ icon are intended to be done on worksheets.

Answers to all workbook exercises are provided as pdf files on the *Media Manager*. The author gratefully acknowledges the careful work of answer writers Professor James H. Andrews of Youngstown State University and Rebecca Sobinovsky.

Acknowledgments: Many thanks to Martha Steele at Addison-Wesley and to Jared Sterzer at WestWords PMG for handling the logistics and production of the *Student Workbook*.

26 Electric Charges and Forces

26.1 Developing a Charge Model

Note: Your answers in Section 26.1 should make *no* mention of electrons or protons.

1. Can an insulator be charged? If so, how would you charge an insulator? If not, why not?

2. Can a conductor be charged? If so, how would you charge a conductor? If not, why not?

3. Lightweight balls A and B hang straight down when both are neutral. They are close enough together to interact, but not close enough to touch. Draw pictures showing how the balls hang if:

   a. Both are touched with a plastic rod that was rubbed with wool.

   b. The two charged balls of part a are moved farther apart.

c. Ball A is touched by a plastic rod that was rubbed with wool and ball B is touched by a glass rod that was rubbed with silk.

d. Both are charged by a plastic rod, but ball A is charged more than ball B.

e. Ball A is charged by a plastic rod. Ball B is neutral.

f. Ball A is charged by a glass rod. Ball B is neutral.

4. Four lightweight balls A, B, C, and D are suspended by threads. Ball A has been touched by a plastic rod that was rubbed with wool. When the balls are brought close together, without touching, the following observations are made:

- Balls B, C, and D are attracted to ball A.
- Balls B and D have no effect on each other.
- Ball B is attracted to ball C.

What are the charge states (glass, plastic, or neutral) of balls A, B, C, and D? Explain.

© 2008 by Pearson Education, Inc., publishing as Pearson Addison-Wesley.

5. Charged plastic and glass rods hang by threads.

   a. An object repels the plastic rod. Can you predict what it will do to the glass rod? If so, what? If not, why not? Explain.

   b. A different object attracts the plastic rod. Can you predict what it will do to the glass rod? If so, what? If not, why not? Explain.

6. After combing your hair briskly, the comb will pick up small pieces of paper.

   a. Is the comb charged? Explain.

   b. How can you be sure that it isn't the paper that is charged? Propose an experiment to test this.

   c. Is your hair charged after being combed? What evidence do you have for your answer?

7. When you take clothes out of the drier right after it stops, the clothes often stick to your hands and arms. Is your body charged? If so, how did it acquire a charge? If not, why does this happen?

8. You've been given a piece of material. Propose an experiment or a series of experiments to determine if the material is a conductor or an insulator. State clearly what the outcome of each experiment will be if the material is a conductor and if it is an insulator.

9. Suppose there exists a third type of charge in addition to the two types we've called glass and plastic. Call this third type X charge. What experiment or series of experiments would you use to test whether an object has X charge? State clearly how each possible outcome of the experiments is to be interpreted.

26.2 Charge

26.3 Insulators and Conductors

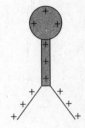

10. A positively charged electroscope has separated leaves.

    a. Suppose you bring a positively charged rod close to the top of the electroscope, but not touching. How will the leaves respond? Use both charge diagrams and words to explain.

    b. How will the leaves respond if you bring a negatively charged rod close to the top of the electroscope, but not touching? Use both charge diagrams and words to explain.

11. a. A negatively charged plastic rod touches a neutral piece of metal. What is the final charge state (positive, negative, or neutral) of the metal? Use both charge diagrams and words to explain how this charge state is achieved.

    b. A positively charged glass rod touches a neutral piece of metal. What is the final charge state of the metal? Use both charge diagrams and words to explain how this charge state is achieved.

12. A lightweight, positively charged ball and a neutral rod hang by threads. They are close but not touching. A positively charged glass rod touches the hanging rod on the end opposite the ball, then the rod is removed.

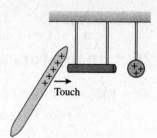

   a. Draw a picture of the final positions of the hanging rod and the ball if the rod is made of glass.

   b. Draw a picture of the final positions of the hanging rod and the ball if the rod is metal.

13. Two oppositely charged metal spheres A and B have equal quantities of charge. They are brought into contact with a neutral metal rod.

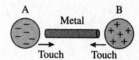

   a. What is the final charge state of each sphere and of the rod?

   b. Give a microscopic explanation, in terms of fundamental charges, of how these final states are reached. Use both charge diagrams and words.

14. Metal sphere A has 4 units of negative charge and metal sphere B has 2 units of positive charge. The two spheres are brought into contact. What is the final charge state of each sphere? Explain.

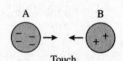

15. a. Metal sphere A is initially neutral. A positively charged rod is brought near, but not touching. Is A now positive, negative, or neutral? Explain.

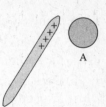

b. Metal spheres A and B are initially neutral and are touching. A positively charged rod is brought near A, but not touching. Is A now positive, negative, or neutral? Explain.

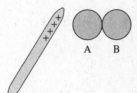

c. Metal sphere A is initially neutral. It is connected by a metal wire to the ground. A positively charged rod is brought near, but not touching. Is A now positive, negative, or neutral? Explain.

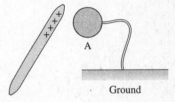

16. A lightweight, positively charged ball and a neutral metal rod hang by threads. They are close but not touching. A positively charged rod is held close to, but not touching, the hanging rod on the end opposite the ball.

a. Draw a picture of the final positions of the hanging rod and the ball. Explain your reasoning.

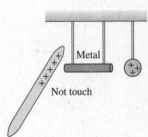

b. Suppose the positively charged rod is replaced with a negatively charged rod. Draw a picture of the final positions of the hanging rod and the ball. Explain your reasoning.

17. A positively charged rod is held near, but not touching, a neutral metal sphere.

    a. Add plusses and minuses to the figure to show the charge distribution on the sphere.

    b. Does the sphere experience a net force? If so, in which direction? Explain.

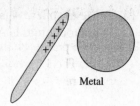

Metal

18. If you bring your finger near a lightweight, negatively charged hanging ball, the ball swings over toward your finger. Use charge diagrams and words to explain this observation.

Finger

19. The figure shows an atom with four protons in the nucleus and four electrons in the electron cloud.

    a. Draw a picture showing how this atom will look if a positive charge is held just *above* the atom.

    b. Is there a net force on the atom? If so, in which direction? Explain.

26.4 Coulomb's Law

20. For each pair of charges, draw a force vector *on each charge* to show the electric force acting on that charge. The length of each vector should be proportional to the magnitude of the force. Each + and − symbol represents the same quantity of charge.

   a.

   b.

   c.

   d.

21. For each group of charges, use a **black** pen or pencil to draw the forces acting on the gray positive charge. Then use a **red** pen or pencil to show the net force on the gray charge. Label $\vec{F}_{net}$.

   a.   b.   c.

22. Can you assign charges (positive or negative) so that these forces are correct? If so, show the charges on the figure. (There may be more than one correct response.) If not, why not?

   a. 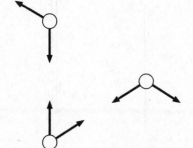   b.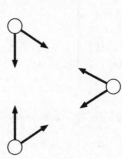

   c.    d.

23. a. Draw a + on the figure below to show the position or positions where a proton would experience no net force.

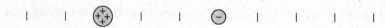

   b. Would the force on an electron at this position (or positions) be to the left, to the right, or zero?

24. Draw a − on the figure below to show the position or positions where an electron would experience no net force.

25. The gray positive charge experiences a net force due to two other charges: the +1 charge that is seen and a +4 charge that is not seen. Add the +4 charge to the figure at the correct position.

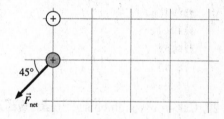

26.5 The Field Model

26. At points 1 to 4, draw an electric field vector with the proper direction and whose length is proportional to the electric field strength at that point.

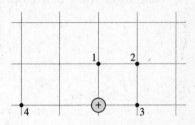

27. Dots 1 and 2 are two points in space. The electric field $\vec{E}_1$ at point 1 is shown. Can you determine the electric field at point 2? If so, draw it on the figure. If not, why not?

28. a. The electric field of a point charge is shown at *one* point in space.

Can you tell if the charge is + or –? If not, why not?

b. Here the electric field of a point charge is shown at two positions in space.

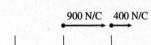

Now can you tell if the charge is + or –? Explain.

c. Can you determine the location of the charge? If so, draw it on the figure. If not, why not?

29. At the three points in space indicated with dots, draw the unit vector $\hat{r}$ that you would use to determine the electric field of the point charge.

    a.

    b.

30. a. This is the unit vector $\hat{r}$ associated with a positive point charge. Draw the electric field vector at this point in space.

    b. This is the unit vector $\hat{r}$ associated with a negative point charge. Draw the electric field vector at this point in space.

31. The electric field strength at a point in space near a point charge is 1000 N/C.

    a. What will be the field strength if the quantity of charge is halved? Explain.

    b. What will be the field strength if the distance to the point charge is halved? The quantity of charge is the original amount, not the value of part a. Explain.

27 The Electric Field

27.1 Electric Field Models

27.2 The Electric Field of Multiple Point Charges

1. You've been assigned the task of determining the magnitude and direction of the electric field at a point in space. Give a step-by-step procedure of how you will do so. List any objects you will use, any measurements you will make, and any calculations you will need to perform. Make sure that your measurements do not disturb the charges that are creating the field.

2. Is there an electric field at the position of the dot? If so, draw the electric field vector on the figure. If not, what would you need to do to create an electric field at this point?

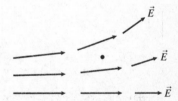

3. This is the electric field in a region of space.
   a. Explain the information that is portrayed in this diagram.

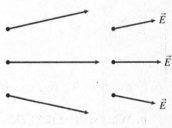

   b. If field vectors were drawn at the same six points but each was only half as long, would the picture represent the same electric field or a different electric field? Explain.

4. Each figure shows two vectors. Can a point charge create an electric field that looks like this at these two points? If so, add the charge to the figure. If not, why not?

Note: The dots are the points to which the vectors are attached. There are no charges at these points.

a.

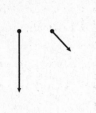

b.

c.

d.

5. At each of the dots, use a **black** pen or pencil to draw and label the electric fields $\vec{E}_1$ and $\vec{E}_2$ due to the two point charges. Make sure that the *relative* lengths of your vectors indicate the strength of each electric field. Then use a **red** pen or pencil to draw and label the net electric field $\vec{E}_{net}$.

a.

b.

6. For each of the figures, use dots to mark any point or points (other than infinity) where $\vec{E} = \vec{0}$.

a.

b.

7. Compare the electric field strengths E_1 and E_2 at the two points labeled 1 and 2. For each, is $E_1 > E_2$, is $E_1 = E_2$, or is $E_1 < E_2$?

a.

⊕ ·1 ⊕ :2

_____

b.

⊕ ·1 ⊖ :2

_____

c.

:1 ⊕ ⊖ :2

_____

d.

2•

⊕ ·1 ⊕

_____

e.

⊕ ·1 ⊕⊕ :2

_____

f.

⊕ ·1 ⊖ :2

_____

8. For each figure, draw and label the net electric field vector $\vec{E}_{net}$ at each of the points marked with a dot or, if appropriate, label the dot $\vec{E}_{net} = \vec{0}$. The lengths of your vectors should indicate the magnitude of $\vec{E}$ at each point.

a.

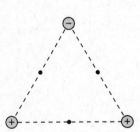

b.

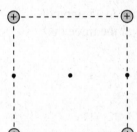

c.

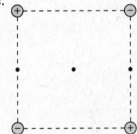

9. At the position of the dot, draw field vectors $\vec{E}_1$ and $\vec{E}_2$ due to q_1 and q_2, and the net electric field $\vec{E}_{net}$. Then, in the blanks, state whether the x- and y-components of $\vec{E}_{net}$ are positive or negative.

a.

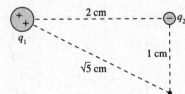

$(E_{net})_x$ _____

$(E_{net})_y$ _____

b.

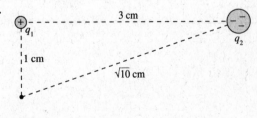

$(E_{net})_x$ _____

$(E_{net})_y$ _____

10. Use a **black** pen or pencil to draw the two electric fields $\vec{E}_1$ and $\vec{E}_2$ at each dot. Then use a **red** pen or pencil to draw $\vec{E}_{net}$. The lengths of your vectors should indicate the magnitude of $\vec{E}$ at each point.

a.   b.

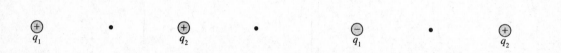

11. Draw the electric field vector at the three points marked with a dot.

    Hint: Think of the charges as horizontal positive/negative pairs, then use superposition.

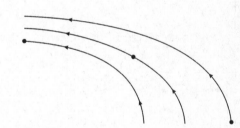

12. The figure shows the electric field lines in a region of space. Draw the electric field vectors at the three dots.

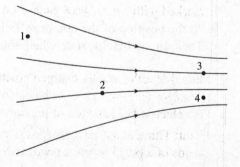

13. The figure shows the electric field lines in a region of space. Rank in order, from largest to smallest, the electric field strengths E_1 to E_4 at points 1 to 4.

    Order:

    Explanation:

© 2008 by Pearson Education, Inc., publishing as Pearson Addison-Wesley.

27.3 The Electric Field of a Continuous Charge Distribution

14. A small segment of wire contains 10 nC of charge.

   a. The segment is shrunk to one-third of its original length. What is the ratio λ_f/λ_i, where λ_i and λ_f are the initial and final linear charge densities?

   b. Suppose the original segment of wire is stretched to 10 times its original length. How much charge must be added to the wire to keep the linear charge density unchanged?

15. A wire has initial linear charge density λ_i. The wire is stretched in length by 50%, and one-third of the charge is removed. What is the ratio λ_f/λ_i, where λ_f is the final linear charge density?

16. The figure shows a uniformly charged positive wire. Five small, equally-spaced segments of charge are shown. Use these five segments to estimate the wire's electric field—both magnitude and direction—at each point in space marked with a dot. Draw each $\vec{E}$ on the figure.

17. Equal-length, equally charged positive and negative wires are placed end-to-end. Draw the electric field at each of the dots.

   Hint: Think about the superposition of the fields of a positive and a negative wire.

18. Two rings of charge face each other. The total charge on each ring is indicated beneath it. Draw the electric field vector on the axis of the rings at the midpoint between them (at the dot), or label the point $\vec{E} = \vec{0}$.

a.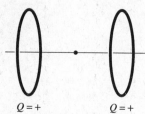

$Q = +$ $Q = +$

b.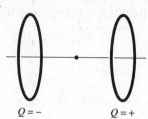

$Q = -$ $Q = +$

c.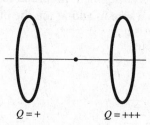

$Q = +$ $Q = +++$

19. The figure shows two charged rods bent into a semicircle. For each, draw the electric field vector at the "center" of the semicircle.

a.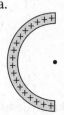

b.

20. A hollow soda straw is uniformly charged. What is the electric field at the center (inside) of the straw? Explain.

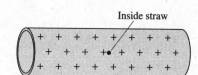

Inside straw

21. An electron experiences a force of magnitude F when it is 1 cm from a very long charged wire with linear charge density λ. If the charge density is doubled, at what distance from the wire will a proton experience a force of the same magnitude F?

27.4 The Electric Fields of Rings, Disks, Planes, and Spheres

22. An irregularly-shaped area of charge has surface charge density η_i.
    Each dimension (x and y) of the area is reduced by a factor of 3.163.

    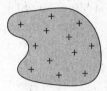

    a. What is the ratio η_f/η_i, where η_f is the final surface charge density?

    b. Compare the final force on a electron very far away to the initial force on the same electron.

23. A circular disk has surface charge density 8 nC/cm^2. What will be the surface charge density if the radius of the disk is doubled?

24. Rank in order, from largest to smallest, the surface charge densities η_1 to η_4 of surfaces 1 to 4.

    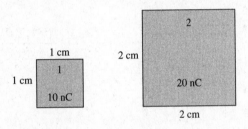
    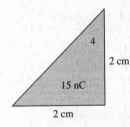

    Order:

    Explanation:

25. A sphere of radius R_i has charge Q_i. What happens to the electric field strength at $r = 2R_i$ if:

    a. The quantity of charge is halved?

    b. The radius of the sphere is halved?

27.5 The Parallel-Plate Capacitor

26. Rank in order, from largest to smallest, the electric field strengths E_1 to E_5 at each of these points.

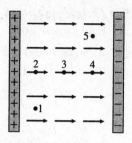

Order:

Explanation:

27. A parallel-plate capacitor is constructed of two square plates, size $L \times L$, separated by distance d. The plates are given charge $\pm Q$. What is the ratio E_f/E_i of the final electric field strength E_f to the initial electric field strength E_i if:

   a. Q is doubled?

   b. L is doubled?

   c. d is doubled?

28. A ball hangs from a thread between two vertical capacitor plates. Initially, the ball hangs straight down. The capacitor plates are charged as shown, then the ball is given a small negative charge. The ball moves to one side, but not enough to touch a capacitor plate.

   a. Draw the ball and thread in the ball's new equilibrium position.
   b. In the space below, draw a free-body diagram of the ball when in its new position.

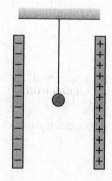

27.6 Motion of a Charged Particle in an Electric Field

27.7 Motion of a Dipole in an Electric Field

29. A small positive charge q experiences a force of magnitude F_1 when placed at point 1. In terms of F_1:

    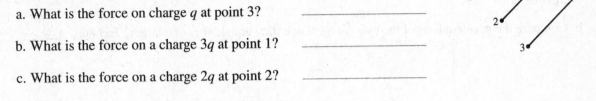

    a. What is the force on charge q at point 3? _____

    b. What is the force on a charge $3q$ at point 1? _____

    c. What is the force on a charge $2q$ at point 2? _____

    d. What is the force on a charge $-2q$ at point 2? _____

30. A small object is released from rest in the center of the capacitor. For each situation, does the object move to the right, to the left, or remain in place? If it moves, does it accelerate or move at constant speed?

    a. Positive object.

    b. Negative object.

    c. Neutral object.

31. Positively and negatively charged objects, with equal masses and equal quantities of charge, enter the capacitor in the directions shown.

    a. Use solid lines to draw their trajectories on the figure if their initial velocities are fast.

    b. Use dashed lines to draw their trajectories on the figure if their initial velocities are slow.

    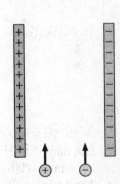

32. An electron is launched from the positive plate at a 45° angle. It does not have sufficient speed to make it to the negative plate. Draw its trajectory on the figure.

    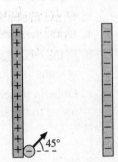

33. A proton and an electron are released from rest in the center of a capacitor.

    a. Compare the forces on the two charges. Are they equal, or is one larger? Explain.

    b. Compare the accelerations of the two charges. Are they equal, or is one larger? Explain.

34. The figure shows an electron orbiting a proton in a hydrogen atom.

    a. What force or forces act on the electron?

    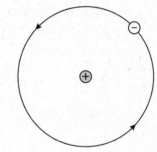

    b. Draw and label the following vectors on the figure: the electron's velocity $\vec{v}$ and acceleration $\vec{a}$, the net force $\vec{F}_{net}$ on the electron, and the electric field $\vec{E}$ at the position of the electron.

35. Does a charged particle always move in the direction of the electric field? If so, explain why. If not, give an example that is otherwise.

36. Three charges are placed at the corners of a triangle. The $++$ charge has twice the quantity of charge of the two $-$ charges; the net charge is zero.

    a. Draw the force vectors on each of the charges.

    b. Is the triangle in equilibrium? _____ If not, draw the equilibrium orientation directly beneath the triangle that is shown.

    c. Once in the equilibrium orientation, will the triangle move to the right, move to the left, rotate steady, or be at rest? Explain.

    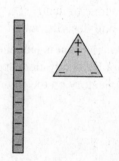

28 Gauss's Law

28.1 Symmetry

1. An infinite plane of charge is seen edge on. The sign of the charge is not given. Do the electric fields shown below have the same symmetry as the charge? If not, why not?

a.

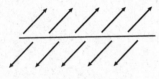

b.

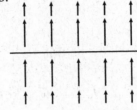

c.

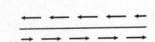

d.

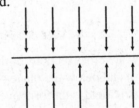

2. Suppose you had a uniformly charged cube. Can you use symmetry alone to deduce the shape of the cube's electric field? If so, sketch and describe the field shape. If not, why not?

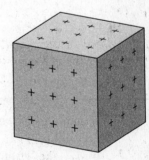

28.2 The Concept of Flux

3. The figures shown below are cross sections of three-dimensional closed surfaces. They have a flat top and bottom surface above and below the plane of the page. However, the electric field is everywhere parallel to the page, so there is no flux through the top or bottom surface. The electric field is uniform over each face of the surface. The field strength, in N/C, is shown.

For each, does the surface enclose a net positive charge, a net negative charge, or no net charge?

a.

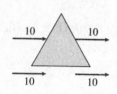

$Q_{net} =$ _____

b.

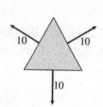

$Q_{net} =$ _____

c.

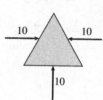

$Q_{net} =$ _____

d.

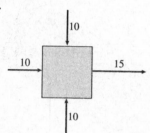

$Q_{net} =$ _____

e.

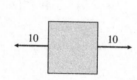

$Q_{net} =$ _____

f.

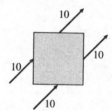

$Q_{net} =$ _____

4. The figures shown below are cross sections of three-dimensional closed surfaces. They have a flat top and bottom surface above and below the plane of the page, but there is no flux through the top or bottom surface. The electric field is uniform over each face of the surface. The field strength, in N/C, is shown.

Each surface contains no net charge. Draw the missing electric field vector (or write $\vec{E} = \vec{0}$) in the proper direction. Write the field strength beside it.

a.

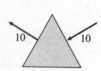

b.

c.

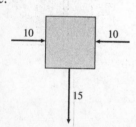

d.

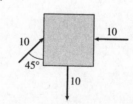

28.3 Calculating Electric Flux

5. Draw the area vector $\vec{A}$ for each of these surfaces.

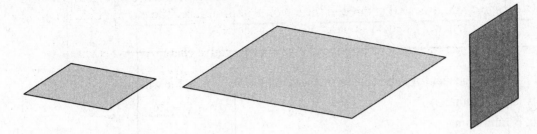

6. How many area vectors are needed to characterize this closed surface?

   Draw them.

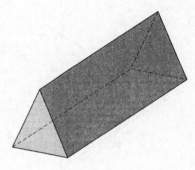

7. The diameter of the circle equals the edge length of the square. Is the electric flux Φ_1 through the square larger than, smaller than, or equal to the electric flux Φ_2 through the circle? Explain.

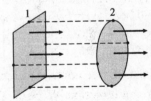

8. Is the electric flux Φ_1 through the circle larger than, smaller than, or equal to the electric flux Φ_2 through the hemisphere? Explain.

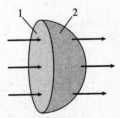

9. A uniform electric field is shown below.

Draw and label an *edge view* of three square surfaces, all the *same size*, for which
a. The flux is maximum.
b. The flux is minimum.
c. The flux has half the value of the flux through square 1.
Give the tilt angle of any squares not perpendicular to the field lines.

10. Is the net electric flux through each of the closed surfaces below positive (+), negative (−), or zero (0)?

a.

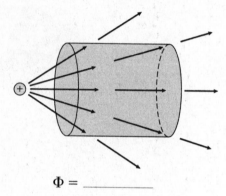

$\Phi =$ _____

b.

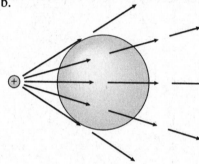

$\Phi =$ _____

c.

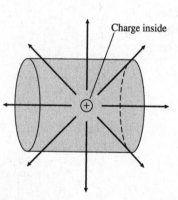

Charge inside

$\Phi =$ _____

d.

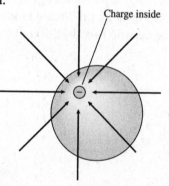

Charge inside

$\Phi =$ _____

e.

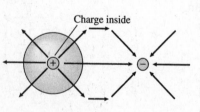

Charge inside

$\Phi =$ _____

f.

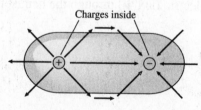

Charges inside

$\Phi =$ _____

28.4 Gauss's Law

28.5 Using Gauss's Law

11. For each of the closed cylinders shown below, are the electric fluxes through the top, the wall, and the bottom positive (+), negative (−), or zero (0)? Is the net flux positive, negative, or zero?

a.

$\Phi_{top} = $ _____

$\Phi_{wall} = $ _____

$\Phi_{bot} = $ _____

$\Phi_{net} = $ _____

b.

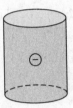

$\Phi_{top} = $ _____

$\Phi_{wall} = $ _____

$\Phi_{bot} = $ _____

$\Phi_{net} = $ _____

c.

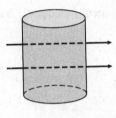

$\Phi_{top} = $ _____

$\Phi_{wall} = $ _____

$\Phi_{bot} = $ _____

$\Phi_{net} = $ _____

d.

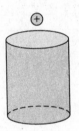

$\Phi_{top} = $ _____

$\Phi_{wall} = $ _____

$\Phi_{bot} = $ _____

$\Phi_{net} = $ _____

e.

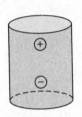

$\Phi_{top} = $ _____

$\Phi_{wall} = $ _____

$\Phi_{bot} = $ _____

$\Phi_{net} = $ _____

f.

$\Phi_{top} = $ _____

$\Phi_{wall} = $ _____

$\Phi_{bot} = $ _____

$\Phi_{net} = $ _____

g.

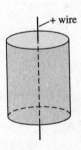

$\Phi_{top} = $ _____

$\Phi_{wall} = $ _____

$\Phi_{bot} = $ _____

$\Phi_{net} = $ _____

h.

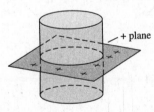

$\Phi_{top} = $ _____

$\Phi_{wall} = $ _____

$\Phi_{bot} = $ _____

$\Phi_{net} = $ _____

i.

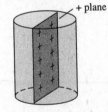

$\Phi_{top} = $ _____

$\Phi_{wall} = $ _____

$\Phi_{bot} = $ _____

$\Phi_{net} = $ _____

12. For this closed cylinder, $\Phi_{top} = -15\ \text{N}\,\text{m}^2/\text{C}$ and $\Phi_{bot} = 5\ \text{N}\,\text{m}^2/\text{C}$. What is Φ_{wall}?

13. What is the electric flux through each of these surfaces? Give your answers as multiples of q/ε_0.

a.

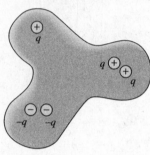

b.

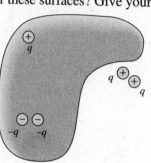

c.

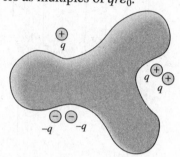

$\Phi_e = $ _____ $\Phi_e = $ _____ $\Phi_e = $ _____

14. What is the electric flux through each of these surfaces? Give your answers as multiples of q/ε_0.

$\Phi_A = $ _____

$\Phi_B = $ _____

$\Phi_C = $ _____

$\Phi_D = $ _____

$\Phi_E = $ _____

15. A charged balloon expands as it is blown up, increasing in size from the initial to final diameters shown. Do the electric fields at points 1, 2, and 3 increase, decrease, or stay the same? Explain your reasoning for each.

Point 1:

Point 2:

Point 3:

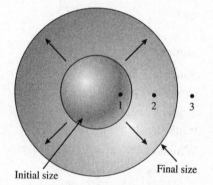

16. Three charges, all the same charge q, are surrounded by three spheres of equal radii.

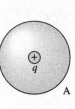

   a. Rank in order, from largest to smallest, the fluxes Φ_1, Φ_2, and Φ_3 through the spheres.

   Order:

   Explanation:

   b. Rank in order, from largest to smallest, the electric field strengths E_1, E_2, and E_3 on the surfaces of the spheres.

   Order:

   Explanation:

17. Two spheres of different diameters surround equal charges. Three students are discussing the situation.

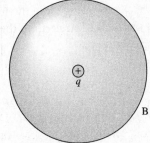

   Student 1: The flux through spheres A and B are equal because they enclose equal charges.

   Student 2: But the electric field on sphere B is weaker than the electric field on sphere A. The flux depends on the electric field strength, so the flux through A is larger than the flux through B.

   Student 3: I thought we learned that flux was about surface area. Sphere B is larger than sphere A, so I think the flux through B is larger than the flux through A.

   Which of these students, if any, do you agree with? Explain.

18. A sphere and an ellipsoid surround equal charges. Four students are discussing the situation.

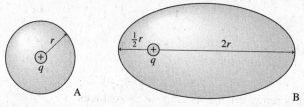

Student 1: The fluxes through A and B are equal because the average radius is the same.

Student 2: I agree that the fluxes are equal, but it's because they enclose equal charges.

Student 3: The electric field is not perpendicular to the surface for B, and that makes the flux through B less than the flux through A.

Student 4: I don't think that Gauss's law even applies to a situation like B, so we can't compare the fluxes through A and B.

Which of these students, if any, do you agree with? Explain.

19. Two parallel, infinite planes of charge have charge densities 2η and $-\eta$. A Gaussian cylinder with cross section A extends distance L to either side.

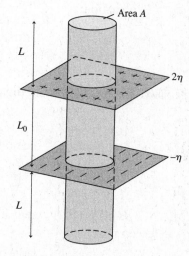

a. Is $\vec{E}$ perpendicular or parallel to the surface at the:

Top _____ Bottom _____ Wall _____

b. Is the electric field E_{top} emerging from the top surface stronger than, weaker than, or equal in strength to the field E_{bot} emerging from the bottom? Explain.

c. By inspection, write the electric fluxes through the three surfaces in terms of E_{top}, E_{bot}, E_{wall}, L, L_0, and A. (You may not need all of these.)

$\Phi_{top} =$ _____ $\Phi_{bot} =$ _____ $\Phi_{wall} =$ _____

d. How much charge is enclosed within the cylinder? Write Q_{in} in terms of η, L, L_0, and A.

$Q_{in} =$ _____

e. By combining your answers from parts b, c, and d, use Gauss's law to determine the electric field strength above the top plane. Show your work.

28.6 Conductors in Electrostatic Equilibrium

20. A small metal sphere hangs by a thread within a larger, hollow conducting sphere. A charged rod is used to transfer positive charge to the outer surface of the hollow sphere.

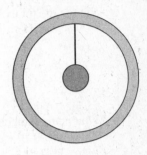

   a. Suppose the thread is an insulator. After the charged rod touches the outer sphere and is removed, are the following surfaces positive, negative, or not charged?

   The small sphere: _____

   The inner surface of the hollow sphere: _____

   The outer surface of the hollow sphere: _____

   b. Suppose the thread is a conductor. After the charged rod touches the outer sphere and is removed, are the following surfaces positive, negative, or not charged?

   The small sphere: _____

   The inner surface of the hollow sphere: _____

   The outer surface of the hollow sphere: _____

21. A small metal sphere hangs by an insulating thread within a larger, hollow conducting sphere. A conducting wire extends from the small sphere through, but not touching, a small hole in the hollow sphere. A charged rod is used to transfer positive charge to the wire. After the charged rod has touched the wire and been removed, are the following surfaces positive, negative, or not charged?

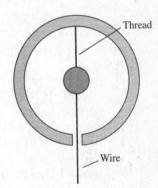

   The small sphere: _____

   The inner surface of the hollow sphere: _____

   The outer surface of the hollow sphere: _____

22. A -10 nC point charge is inside a hole in a conductor. The conductor has no net charge.

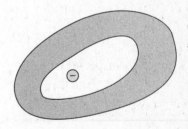

   a. What is the total charge on the inside surface of the conductor?

   b. What is the total charge on the outside surface of the conductor?

23. A -10 nC point charge is inside a hole in a conductor. The conductor has a net charge of $+10$ nC.

   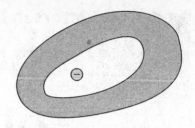

   a. What is the total charge on the inside surface of the conductor?

   b. What is the total charge on the outside surface of the conductor?

24. An insulating thread is used to lower a positively charged metal ball into a metal container. Initially, the container has no net charge. Use plus and minus signs to show the charge distribution on the ball at the times shown in the figure. (The ball's charge is already shown in the first frame.)

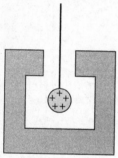

Ball hasn't touched

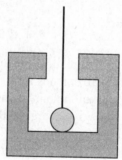

Ball has touched

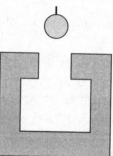

Ball has been withdrawn

29 The Electric Potential

29.1 Electric Potential Energy

29.2 The Potential Energy of Point Charges

1. A positive point charge and a negative point charge are inside a parallel-plate capacitor. The point charges interact only with the capacitor, not with each other. Let the negative capacitor plate be the zero of potential energy for both charges.

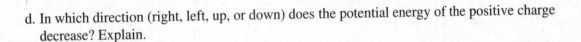

   a. Use a **black** pen or pencil to draw the electric field vectors inside the capacitor.

   b. Use a **red** pen or pencil to draw the forces acting on the two charges.

   c. Is the potential energy of the *positive* point charge positive, negative, or zero? Explain.

   d. In which direction (right, left, up, or down) does the potential energy of the positive charge decrease? Explain.

   e. In which direction will the positive charge move if released from rest? Use the concept of energy to explain your answer.

   f. Does your answer to part e agree with the force vector you drew in part b? _____

   g. Repeat steps c to f for the *negative* point charge.

2. A positive charge q is fired through a small hole in the positive plate of a capacitor. Does q speed up or slow down inside the capacitor? Answer this question twice:

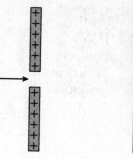

a. First using the concept of force.

b. Second using the concept of energy.

3. Charge $q_1 = 3$ nC is distance r from a positive point charge Q. Charge $q_2 = 1$ nC is distance $2r$ from Q. What is the ratio U_1/U_2 of their potential energies due to their interactions with Q?

4. The figure shows the potential energy of a positively charged particle in a region of space.

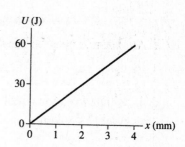

a. What possible arrangement of source charges is responsible for this potential energy? Draw the source charges above the axis below.

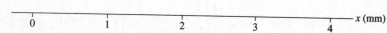

b. With what kinetic energy should the charged particle be launched from $x = 0$ mm to have a turning point at $x = 3$ mm? Explain.

c. How much kinetic energy does this charged particle of part b have as it passes $x = 2$ mm?

5. An electron ($q = -e$) completes half of a circular orbit of radius r around a nucleus with $Q = +3e$.

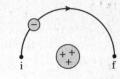

   a. How much work is done on the electron as it moves from i to f? Give either a numerical value or an expression from which you could calculate the value if you knew the radius. Justify your answer.

   b. By how much does the electric potential energy change as the electron moves from i to f?

   c. Is the electron's speed at f greater than, less than, or equal to its speed at i?

   d. Are your answers to parts a and c consistent with each other?

6. An electron moves along the trajectory from i to f.

   a. Does the electric potential energy increase, decrease, or stay the same? Explain.

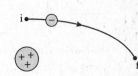

   b. Is the electron's speed at f greater than, less than, or equal to its speed at i? Explain.

7. Inside a parallel-plate capacitor, two protons are launched with the same speed from point 1. One proton moves along the path from 1 to 2, the other from 1 to 3. Points 2 and 3 are the same distance from the positive plate.

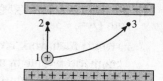

a. Is $\Delta U_{1 \to 2}$, the change in potential energy along the path $1 \to 2$, larger than, smaller than, or equal to $\Delta U_{1 \to 3}$? Explain.

b. Is the proton's speed v_2 at point 2 larger than, smaller than, or equal to v_3? Explain.

29.3 The Potential Energy of a Dipole

8. Rank in order, from most positive to most negative, the potential energies U_1 to U_6 of these six electric dipoles in a uniform electric field.

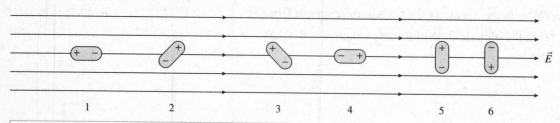

Order:

Explanation:

29.4 The Electric Potential

9. Charged particles with $q = +0.1$ C are fired with 10 J of kinetic energy toward a region of space in which there is an electric potential. The figure shows the kinetic energy of the charged particles as they arrive at nine different points in the region. Determine the electric potential at each of these points. Write the value of the potential above each of the dots. Assume that the particles start from a point where the electric potential is zero.

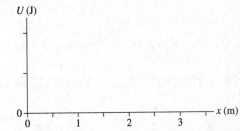

8 J   6 J   5 J   6 J   8 J

7 J   6 J   7 J

7 J

10. a. The graph on the left shows the electric potential along the x-axis. Use the axes on the right to draw a graph of the potential energy of a 0.1 C charged particle in this region of space. Provide a numerical scale on the energy axis.

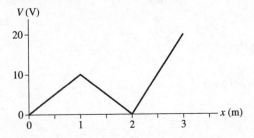

b. If the charged particle is shot toward the right from $x = 1$ m with 1.0 J of kinetic energy, where is its turning point? Explain.

c. Will the charged particle of part b ever reach $x = 0$ m? If so, how much kinetic energy will it have at that point? If not, why not?

29.5 The Electric Potential Inside a Parallel-Plate Capacitor

11. Rank in order, from largest to smallest, the electric potentials V_1 to V_5 at points 1 to 5.

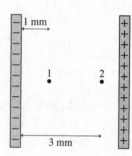

Order:

Explanation:

12. The figure shows two points inside a capacitor. Let $V = 0$ V at the negative plate.

    a. What is the ratio V_2/V_1 of the electric potential at these two points? Explain.

    b. What is the ratio E_2/E_1 of the electric field strength at these two points? Explain.

13. The figure shows two capacitors, each with a 3 mm separation. A proton is released from rest in the center of each capacitor.

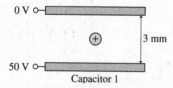

Capacitor 1

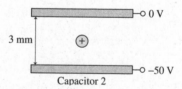

Capacitor 2

    a. Draw an arrow on each proton to show the direction it moves.

    b. Which proton reaches a capacitor plate first? Or are they simultaneous? Explain.

14. A capacitor with plates separated by distance d is charged to a potential difference ΔV_C. All wires and batteries are disconnected, then the two plates are pulled apart (with insulated handles) to a new separation of distance $2d$.

    a. Does the capacitor charge Q change as the separation increases? If so, by what factor? If not, why not?

    b. Does the electric field strength E change as the separation increases? If so, by what factor? If not, why not?

    c. Does the potential difference ΔV_C change as the separation increases? If so, by what factor? If not, why not?

15. Each figure shows a contour map on the left and a set of graph axes on the right. Draw a graph of V versus x. Your graph should be a straight line or a smooth curve.

    a.

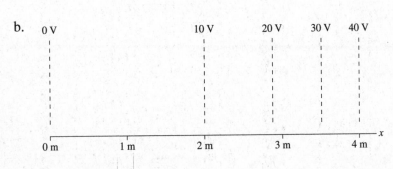
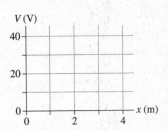

    b.

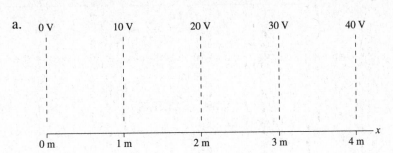
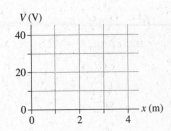

c.
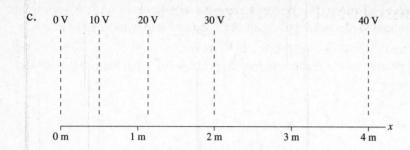
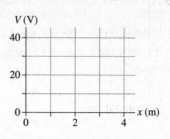

16. Each figure shows a *V*-versus-*x* graph on the left and an *x*-axis on the right. Assume that the potential varies with *x* but not with *y*. Draw a contour map of the electric potential. Your figures should look similar to the contour maps in Question 15. There should be a uniform difference between equipotential lines, and each equipotential line should be labeled.

a.
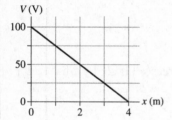

b.
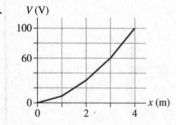

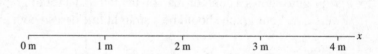

29.6 The Electric Potential of a Point Charge

17. Rank in order, from largest to smallest, the electric potentials V_1 to V_5 at points 1 to 5.

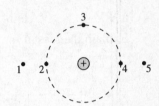

Order:

Explanation:

18. Rank in order, from least negative to most negative, the electric potentials V_1 to V_5 at points 1 to 5.

Order:

Explanation:

19. The figure shows two points near a positive point charge.

    a. What is the ratio V_1/V_2 of the electric potentials at these two points? Explain.

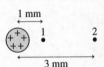

    b. What is the ratio E_1/E_2 of the electric field strengths at these two points? Explain.

20. A 1 nC positive point charge is located at point A. The electric potential
    at point B is

    a. 9 V   b. $9 \cdot \sin 30°$ V   c. $9 \cdot \cos 30°$ V   d. $9 \cdot \tan 30°$ V

    Explain the reason for your choice.

21. An inflatable metal balloon of radius R is charged to a potential of 1000 V. After all wires and batteries
    are disconnected, the balloon is inflated to a new radius $2R$.

    a. Does the potential of the balloon change as it is inflated? If so, by what factor? If not, why not?

    b. Does the potential at a point at distance $r = 4R$ change as the balloon is inflated? If so, by what
       factor? If not, why not?

29.7 The Electric Potential of Many Charges

22. Each figure below shows three points in the vicinity of two point charges. The charges have equal magnitudes. Rank in order, from largest to smallest, the potentials V_1, V_2, and V_3.

a.

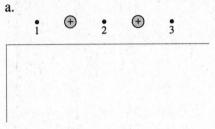

b.

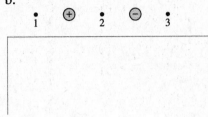

c.

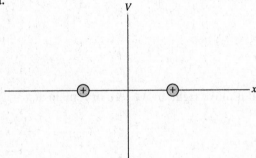

d.

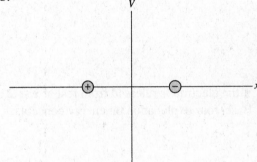

23. On the axes below, draw a graph of V versus x for the two point charges shown.

a.

b.

24. For each pair of charges below, are there any points (other than at infinity) at which the electric potential is zero? If so, identify them on the figure with a dot and a label. If not, why not?

a.

b.

25. For each pair of charges below, at which grid point or points could a double-negative point charge ($q = -2$) be placed so that the potential at the dot is 0 V? There may be more than one possible point. Draw the charge on the figure at all points that work.

a.

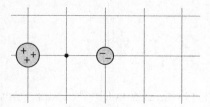

b.

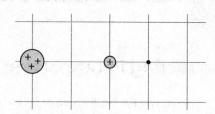

26. The graph shows the electric potential along the x-axis due to point charges on the x-axis.

   a. Draw the charges *on the axis of the figure*. Note that the charges may have different magnitudes.

   b. An electron is placed at $x = 2$ cm. Is its potential energy positive, negative, or zero? Explain.

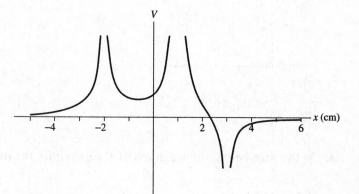

   c. If the electron is released from rest at $x = 2$ cm, will it move right, move left, or remain at $x = 2$ cm? Base your explanation on energy concepts.

27. A ring has radius R and charge Q. The ring is shrunk to a new radius $\frac{1}{2}R$ with no change in its charge. By what factor does the on-axis potential at $z = R$ increase?

30 Potential and Field

30.1 Connecting Potential and Field

1. The top graph shows the *x*-component of $\vec{E}$ as a function of *x*. On the axes below the graph, draw the graph of *V* versus *x* in this same region of space. Let $V = 0$ V at $x = 0$ m. Include an appropriate vertical scale. (Hint: Integration is the area under the curve.)

a.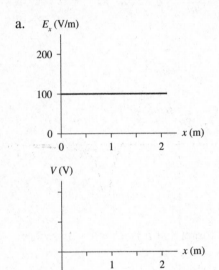

b.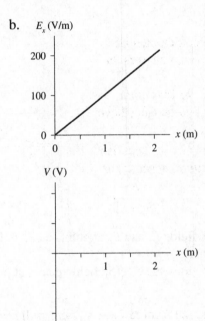

30.2 Sources of Electric Potential

2. What is ΔV_{series} for each group of 1.5 V batteries?

a.  $\Delta V_{series} =$ _____

b. $\Delta V_{series} =$ _____

c. $\Delta V_{series} =$ _____

30.3 Finding the Electric Field from the Potential

3. The top graph shows the electric potential as a function of x. On the axes below the graph, draw the graph of E_x versus x in this same region of space. Add an appropriate scale on the vertical axis.

a.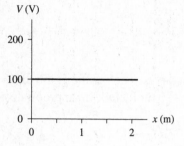

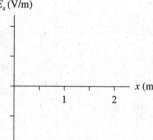

b.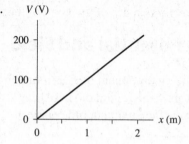

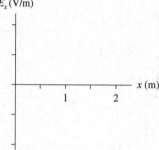

4. For each contour map:
   i. Estimate the electric fields $\vec{E}_a$ and $\vec{E}_b$ at points a and b. Don't forget that $\vec{E}$ is a vector. Show how you made your estimate.
   ii. On the contour map, draw the electric field vectors at points a and b.

a.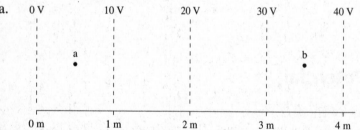

$\vec{E}_a =$ _____

$\vec{E}_b =$ _____

b.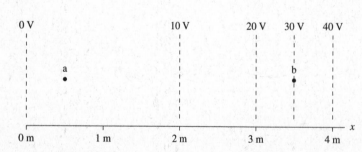

$\vec{E}_a =$ _____

$\vec{E}_b =$ _____

5. The top graph shows E_x versus x for an electric field that is parallel to the x-axis.
   a. Draw the graph of V versus x in this region of space. Let $V = 0$ V at $x = 0$ m. Add an appropriate scale on the vertical axis. (Hint: Integration is the area under the curve.)
   b. Use dashed lines to draw a contour map above the x-axis on the right. Space your equipotential lines every 20 volts and label each equipotential line.
   c. Draw electric field vectors on top of the contour map.

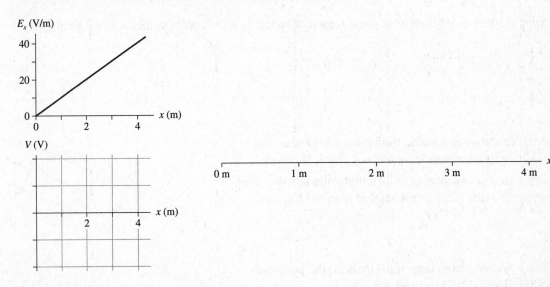

6. Draw the electric field vectors at the dots on this contour map. The length of each vector should be proportional to the field strength at that point.

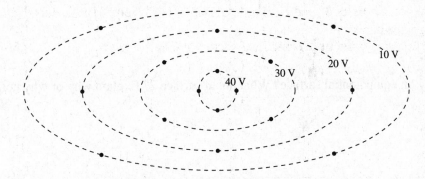

7. Draw the electric field vectors at the dots on this contour map. The length of each vector should be proportional to the field strength at that point.

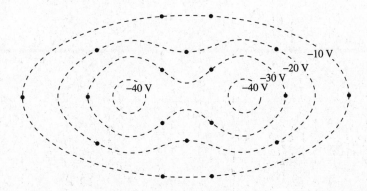

8. a. Suppose $\vec{E} = \vec{0}$ V/m throughout some region of space. Is $V = 0$ V in this region? Explain.

b. Suppose $V = 0$ V throughout some region of space. Is $\vec{E} = \vec{0}$ V/m in this region? Explain.

9. The figure shows an electric field diagram. Dashed lines 1 and 2 are two surfaces in space, not physical objects.

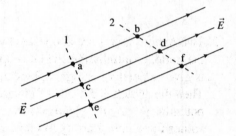

   a. Is the electric potential at point a higher than, lower than, or equal to the electric potential at point b? Explain.

   b. Rank in order, from largest to smallest, the potential differences ΔV_{ab}, ΔV_{cd}, and ΔV_{ef}.

   Order:

   Explanation:

   c. Is surface 1 an equipotential surface? What about surface 2? Explain why or why not.

10. For each of the figures below, is this a physically possible potential map if there are no free charges in this region of space? If so, draw an electric field line diagram on top of the potential map. If not, why not?

   a.
   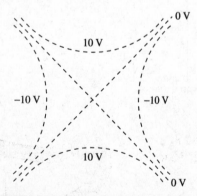

   b.
   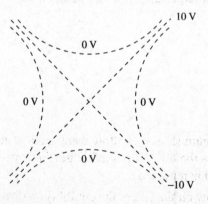

30.4 A Conductor in Electrostatic Equilibrium

11. The figure shows a negatively charged electroscope. The gold leaf stands away from the rigid metal post. Is the electric potential of the leaf higher than, lower than, or equal to the potential of the post? Explain.

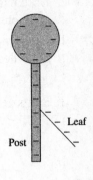

12. Two metal spheres are connected by a metal wire that has a switch in the middle. Initially the switch is open. Sphere 1, with the larger radius, is given a positive charge. Sphere 2, with the smaller radius, is neutral. Then the switch is closed. Afterward, sphere 1 has charge Q_1, is at potential V_1, and the electric field strength at its surface is E_1. The values for sphere 2 are Q_2, V_2, and E_2.

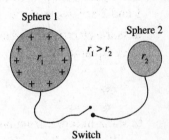

a. Is V_1 larger than, smaller than, or equal to V_2? Explain.

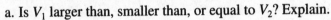

b. Is Q_1 larger than, smaller than, or equal to Q_2? Explain.

c. Is E_1 larger than, smaller than, or equal to E_2? Explain.

13. The figure shows a hollow metal shell. A negatively charged rod touches the top of the sphere, transferring charge to the shell. Then the rod is removed.

    a. Show on the figure the equilibrium distribution of charge.

    b. Draw the electric field diagram.

14. The figure shows two flat metal electrodes that are held at potentials of 100 V and 0 V.

   a. Used dashed lines to sketch a reasonable approximation of the 20 V, 40 V, 60 V, and 80 V equipotential lines.

   b. Draw enough electric field lines to indicate the shape of the electric field. Use solid lines with arrowheads.

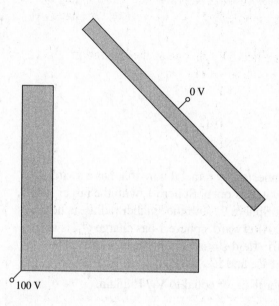

15. The figure shows two 3 V batteries with metal wires attached to each end. Points a and c are *inside* the wire. Point b is inside the battery. For each figure:

- What are the potential differences ΔV_{12}, ΔV_{23}, ΔV_{34}, and ΔV_{14}?
- Does the electric field at a, b, and c point left, right, up, or down? Or is $\vec{E} = \vec{0}$?

a.

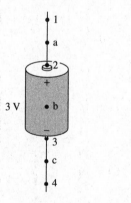

b.

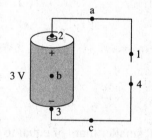

$\Delta V_{12} = \underline{\hspace{1cm}}$ $\Delta V_{23} = \underline{\hspace{1cm}}$

$\Delta V_{34} = \underline{\hspace{1cm}}$ $\Delta V_{14} = \underline{\hspace{1cm}}$

$\vec{E}_a \underline{\hspace{1cm}}$ $\vec{E}_b \underline{\hspace{1cm}}$ $\vec{E}_c \underline{\hspace{1cm}}$

$\Delta V_{12} = \underline{\hspace{1cm}}$ $\Delta V_{23} = \underline{\hspace{1cm}}$

$\Delta V_{34} = \underline{\hspace{1cm}}$ $\Delta V_{14} = \underline{\hspace{1cm}}$

$\vec{E}_a \underline{\hspace{1cm}}$ $\vec{E}_b \underline{\hspace{1cm}}$ $\vec{E}_c \underline{\hspace{1cm}}$

30.5 Capacitance and Capacitors

30.6 The Energy Stored in a Capacitor

16. A parallel-plate capacitor with plate separation d is connected to a battery that has potential difference ΔV_{bat}. Without breaking any of the connections, insulating handles are used to increase the plate separation to $2d$.

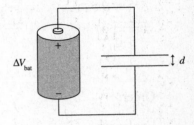

a. Does the potential difference ΔV_C change as the separation increases? If so, by what factor? If not, why not?

b. Does the capacitance change? If so, by what factor? If not, why not?

c. Does the capacitor charge Q change? If so, by what factor? If not, why not?

17. For the capacitor shown, the potential difference ΔV_{ab} between points a and b is

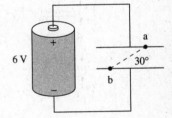

a. 6 V
b. $6 \cdot \sin 30°$ V
c. $6/\sin 30°$ V
d. $6 \cdot \tan 30°$ V
e. $6 \cdot \cos 30°$ V
f. $6/\cos 30°$ V

Explain your choice.

18. Rank in order, from largest to smallest, the potential differences $(\Delta V_C)_1$ to $(\Delta V_C)_4$ of these four capacitors.

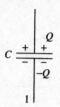 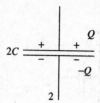 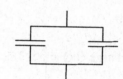

Order:

Explanation:

19. Each capacitor in the circuits below has capacitance C. What is the equivalent capacitance of the group of capacitors?

a.

$C_{eq} =$ _____

b.

$C_{eq} =$ _____

c.

$C_{eq} =$ _____

d.

$C_{eq} =$ _____

e.

$C_{eq} =$ _____

f.

$C_{eq} =$ _____

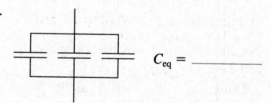

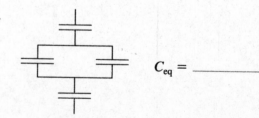

20. Rank in order, from largest to smallest, the equivalent capacitances $(C_{eq})_1$ to $(C_{eq})_4$ of these four groups of capacitors.

 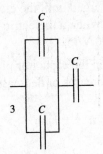 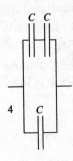 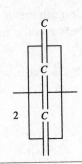

Order:

Explanation:

21. Rank in order, from largest to smallest, the energies $(U_C)_1$ to $(U_C)_4$ stored in each of these capacitors.

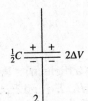 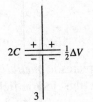 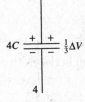

Order:

Explanation:

30.7 Dielectrics

22. An air-insulated capacitor is charged until the electric field strength inside is 10,000 V/m, then disconnected from the battery. When a dielectric is inserted between the capacitor plates, the electric field strength is reduced to 2000 V/m.

   a. Does the amount of charge on the capacitor plates increase, decrease, or stay the same when the dielectric is inserted? If it increases or decreases, by what factor?

   b. Does the potential difference between the capacitor plates increase, decrease, or stay the same when the dielectric is inserted? If it increases or decreases, by what factor?

23. The plates of an air-insulated capacitor are charged to ±100 nC, then left connected to the battery. When a dielectric is inserted between the plates, the charge on the plates increases to ±500 nC.

   a. Does the potential difference across the capacitor increase, decrease, or stay the same when the dielectric is inserted? If it increases, by what factor?

   b. Does the electric field strength inside the capacitor increase, decrease, or stay the same when the dielectric is inserted? If it increases, by what factor?

24. The gap between two capacitor plates is *partially* filled with a dielectric. Rank in order, from largest to smallest, the electric field strengths E_1, E_2, and E_3 at points 1, 2, and 3.

   Order:

   Explanation:

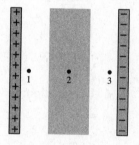

31 Current and Resistance

31.1 The Electron Current

1. A lightbulb is connected with wires to a battery, and the bulb is glowing. Are simple observations and measurements you can make on this circuit able to distinguish a current composed of positive charge carriers from a current composed of negative charge carriers? If so, describe how you can tell which it is. If not, why not?

2. Describe experimental evidence to support the claim that the charge carriers in metals are electrons. Use both pictures and words.

3. Are the charge carriers always electrons? If so, why is this the case? If not, describe a situation in which a current is due to some other charge carrier.

31.2 Creating a Current

4. The electron drift speed in a wire is exceedingly slow—typically only a fraction of a millimeter per second. Yet when you turn on a light switch, a light bulb several meters away seems to come on instantly. Explain how to resolve this apparent paradox.

5. The figure shows a segment of a current-carrying metal wire.

   a. Is there an electric field inside the wire? If so, draw and label an arrow on the figure to show its direction. If not, why not?

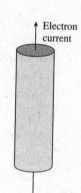

Electron current

   b. If there is an electric field, draw on the figure a possible arrangement of charges that could be the source charges causing the field.

6. a. If the electrons in a current-carrying wire collide with the positive ions *more* frequently, does their drift speed increase or decrease? Explain.

   b. Does an increase in the collision frequency make the wire a better conductor or a worse conductor? Explain.

   c. Would you expect a metal to be a better conductor at high temperature or at low temperature? Explain.

31.3 Current and Current Density

7. What is the difference between current and current density?

8. The figure shows a segment of a current-carrying metal wire.

    a. Draw an arrow on the figure, using a **black** pen or pencil, to show the direction of motion of the charge carriers.

    b. Draw an arrow on the figure, using a **red** pen or pencil, to show the direction of the electric field.

9. Is I_2 greater than, less than, or equal to I_1? Explain.

10. All wires in this figure are made of the same material and have the same diameter. Rank in order, from largest to smallest, the currents I_1 to I_4.

    Order:

    Explanation:

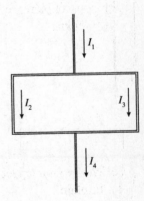

11. A lightbulb is connected to a battery with 1-mm-diameter
    wires. The bulb is glowing.

    a. Draw arrows at points 1, 2, and 3 to show the direction of
    the electric field at those points. (The points are *inside* the
    wire.)

    b. Rank in order, from largest to smallest, the field strengths
    E_1, E_2, and E_3.

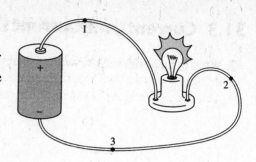

Order:

Explanation:

12. A wire carries a 4 A current. What is the current in a second wire that delivers twice as much charge in
    half the time?

13. The current density in a wire is 1000 A/m². What will the current density be if the current is doubled
    and the wire's diameter is halved?

31.4 Conductivity and Resistivity

14. Metal 1 and metal 2 are each formed into 1-mm-diameter wires. The electric field needed to cause a 1 A current in metal 1 is larger than the electric field needed to cause a 1 A current in metal 2. Which metal has the larger conductivity? Explain.

15. If a metal is heated, does its conductivity increase, decrease, or stay the same? Explain.

16. Wire 1 and wire 2 are made from the same metal. Wire 1 has twice the diameter and half the electric field of wire 2. What is the ratio I_1/I_2?

17. Wire 1 and wire 2 are made from the same metal. Wire 2 has a larger diameter than wire 1. The electric field strengths E_1 and E_2 are equal.

   a. Compare the values of the two current densities. Is J_1 greater than, less than, or equal to J_2? Explain.

   b. Compare the values of the currents I_1 and I_2.

   c. Compare the values of the electron drift speeds $(v_d)_1$ and $(v_d)_2$.

18. A wire consists of two segments of different diameters but made from the same metal. The current in segment 1 is I_1.

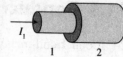

   a. Compare the values of the currents in the two segments. Is I_2 greater than, less than, or equal to I_1? Explain.

   b. Compare the values of the current densities J_1 and J_2.

   c. Compare the strengths of the electric fields E_1 and E_2 in the two segments.

   d. Compare the values of the electron drift speeds $(v_d)_1$ and $(v_d)_2$.

19. A wire consists of two equal-diameter segments. Their conductivities and electron densities differ, with $\sigma_2 > \sigma_1$ and $n_2 > n_1$. The current in segment 1 is I_1.

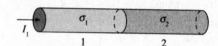

   a. Compare the values of the currents in the two segments. Is I_2 greater than, less than, or equal to I_1? Explain.

   b. Compare the strengths of the current densities J_1 and J_2.

   c. Compare the strengths of the electric fields E_1 and E_2 in the two segments.

   d. Compare the values of the electron drift speeds $(v_d)_1$ and $(v_d)_2$.

31.5 Resistance and Ohm's Law

20. A continuous metal wire connects the two ends of a 3 V battery with a rectangular loop. The negative terminal of the battery has been chosen as the point where $V = 0$ V.

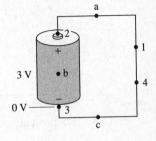

   a. Locate and label the approximate points along the wire where $V = 3$ V, $V = 2$ V, and $V = 1$ V.

   b. Points a and c are *inside* the wire. Point b is inside the battery. Does the electric field at a, b, and c point left, right, up, or down? Or is $\vec{E} = \vec{0}$?

   $\vec{E}_a$ —————— $\vec{E}_b$ —————— $\vec{E}_c$ ——————

   c. In moving through the *wire* from point 2 to point 3, does the potential increase, decrease, or not change? If the potential changes, by how much does it change?

   d. In moving through the *battery* from point 2 to point 3, does the potential increase, decrease, or not change? If the potential changes, by how much does it change?

   e. In moving all the way around the loop in a clockwise direction, starting from point 2 and ending at point 2, is the net change in the potential positive, negative, or zero?

21. a. Which direction—clockwise or counterclockwise—does an electron travel through the wire? Explain.

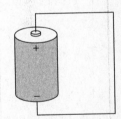

   b. Does an electron's electric potential energy increase, decrease, or stay the same as it moves through the wire? Explain.

   c. If you answered "decrease" in part b, where does the energy go? If you answered "increase" in part b, where does the energy come from?

22. The wires below are all made of the same material. Rank in order, from largest to smallest, the resistances R_1 to R_5 of these wires.

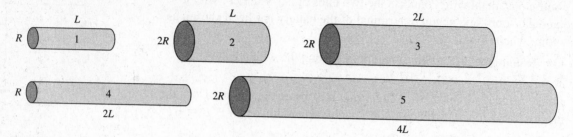

Order:

Explanation:

23. The two circuits use identical batteries and wires of equal diameters. Rank in order, from largest to smallest, the currents I_1, I_2, I_3, and I_4 at points 1 to 4.

Order:

Explanation:

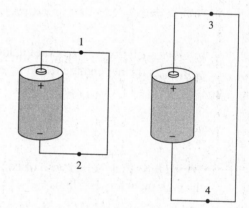

24. The two circuits use identical batteries and wires of equal diameters. Rank in order, from largest to smallest, the currents I_1 to I_7 at points 1 to 7.

Order:

Explanation:

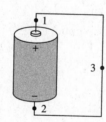
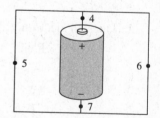

25. A wire is connected to the terminals of a 6 V battery. What is the potential difference ΔV_{wire} between the ends of the wire, and what is the current I through the wire, if the wire has the following resistances:

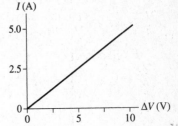

6 V

Wire resistance R

a. $R = 1\,\Omega$ $\Delta V_{wire} = $ _____ $I = $ _____

b. $R = 2\,\Omega$ $\Delta V_{wire} = $ _____ $I = $ _____

c. $R = 3\,\Omega$ $\Delta V_{wire} = $ _____ $I = $ _____

d. $R = 6\,\Omega$ $\Delta V_{wire} = $ _____ $I = $ _____

26. The graph shows the current-versus-potential-difference relationship for a resistor R.

    a. What is the numerical value of R?

I (A)

5.0

2.5

0

0   5   10

ΔV (V)

   b. Suppose the length of the resistor is doubled. On the figure, draw the current-versus-potential-difference graph for the longer resistor.

27. For resistors R_1 and R_2:

    a. Which end (left, right, top, or bottom) is more positive?

    R_1 _____ R_2 _____

    b. In which direction (such as left to right or top to bottom) does the potential decrease?

    R_1 _____

    R_2 _____

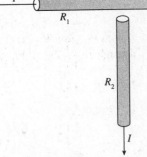

I

R_1

R_2

I

28. Rank in order, from largest to smallest, the currents I_1 to I_4 through these four resistors.

+ 2 V –   2 Ω I_1 + 1 V –   2 Ω I_2 + 2 V –   1 Ω I_3 + 1 V –   1 Ω I_4

Order:

Explanation:

29. Which, if any, of these statements are true? (More than one may be true.)

    i.   A battery supplies the energy to a circuit.

    ii. A battery is a source of potential difference. The potential difference between the terminals of the battery is always the same.

    iii. A battery is a source of current. The current leaving the battery is always the same.

    Explain your choice or choices.

32 Fundamentals of Circuits

32.1 Circuit Elements and Diagrams

32.2 Kirchhoff's Laws and the Basic Circuit

1. The tip of a flashlight bulb is touching the top of a 3 V battery. Does the bulb light? Why or why not?

2. Current I_{in} flows into three resistors connected together one after the other. The graph shows the value of the potential as a function of distance.

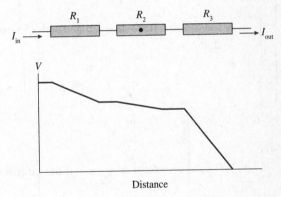

a. Is I_{out} greater than, less than, or equal to I_{in}? Explain.

b. Rank in order, from largest to smallest, the three resistances R_1, R_2, and R_3.

Order:

Explanation:

c. Is there an electric field at the point inside R_2 that is marked with a dot? If so, in which direction does it point? If not, why not?

3. A flashlight bulb is connected to a battery and is glowing. Is current I_2 greater than, less than, or equal to current I_1? Explain.

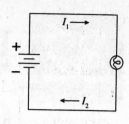

4. a. In which direction does current flow through resistor R?

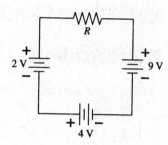

   b. Which end of R is more positive? Explain.

   c. If this circuit were analyzed in a clockwise direction, what numerical value would you assign to ΔV_R? Why?

   d. What value would ΔV_R have if the circuit were analyzed in a counterclockwise direction?

5. The wire is broken on the right side of this circuit. What is the potential difference ΔV_{12} between points 1 and 2? Explain.

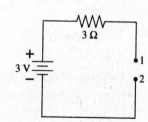

6. Draw a circuit for which the Kirchhoff loop law equation is

$$6V - I \cdot 2\Omega + 3V - I \cdot 4\Omega = 0$$

Assume that the analysis is done in a clockwise direction.

7. The current in a circuit is 2.0 A. The graph shows how the potential changes when going around the circuit in a clockwise direction, starting from the lower left corner. Draw the circuit diagram.

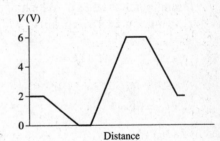

32.3 Energy and Power

8. This circuit has two resistors, with $R_1 > R_2$. Which of the two resistors dissipates the larger amount of power? Explain.

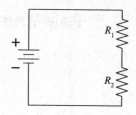

9. Two conductors of equal lengths are connected to a battery by ideal wires. The conductors are made of the same material but have different radii. Which of the two conductors dissipates the larger amount of power? Explain.

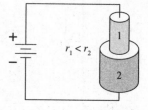

10. Two conductors of equal lengths are connected to a battery by ideal wires. The conductors have the same radii but are made of different materials and have different conductivities σ. Which of the two conductors dissipates the larger amount of power? Explain.

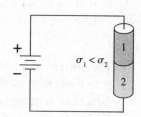

11. A 60 W lightbulb and a 100 W lightbulb are placed one after the other in a circuit. The battery's emf is large enough that both bulbs are glowing. Which one glows more brightly? Explain.

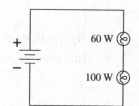

32.4 Series Resistors

32.5 Real Batteries

12. What is the equivalent resistance of each group of resistors?

a.   b.   c.

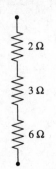

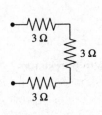

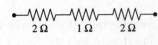

$R_{eq} =$ _____ $R_{eq} =$ _____ $R_{eq} =$ _____

13. The figure shows two circuits. The two batteries are identical and the four resistors all have exactly the same resistance.

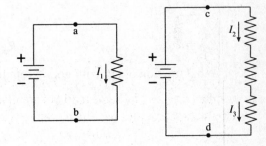

   a. Is ΔV_{ab} larger than, smaller than, or equal to ΔV_{cd}? Explain.

   b. Rank in order, from largest to smallest, the currents I_1, I_2, and I_3.

   Order:

   Explanation:

14. The lightbulb in this circuit has a resistance of $1\,\Omega$.

   a. What are the values of:

   ΔV_{12} _____

   ΔV_{23} _____

   ΔV_{34} _____

   b. Suppose the bulb is now removed from its socket. Then what are the values of:

   ΔV_{12} _____

   ΔV_{23} _____

   ΔV_{34} _____

15. If the value of R is increased, does ΔV_{bat} increase, decrease, or stay the same? Explain.

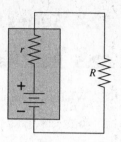

32.6 Parallel Resistors

16. What is the equivalent resistance of each group of resistors?

a.

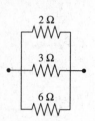

$R_{eq} = $ _____

b.

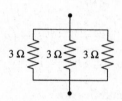

$R_{eq} = $ _____

c.

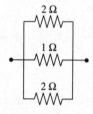

$R_{eq} = $ _____

17. a. What fraction of current I goes through the $3\,\Omega$ resistor?

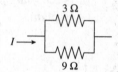

b. If the $9\,\Omega$ resistor is replaced with a larger resistor, will the fraction of current going through the $3\,\Omega$ resistor increase, decrease, or stay the same?

18. The figure shows five combinations of identical resistors. Rank in order, from largest to smallest, the equivalent resistances $(R_{eq})_1$ to $(R_{eq})_5$.

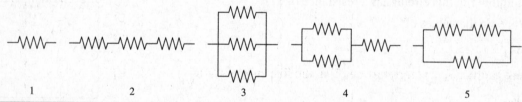

Order:

Explanation:

32.7 Resistor Circuits

32.8 Getting Grounded

19. The circuit shown has a battery and two resistors, with $R_1 > R_2$. Which of the two resistors dissipates the larger amount of power? Explain your reasoning.

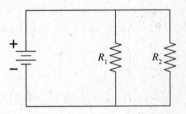

20. Rank in order, from largest to smallest, the three currents I_1 to I_3.

    Order:

    Explanation:

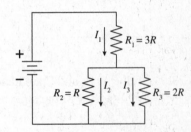

21. The two batteries are identical and the four resistors all have exactly the same resistance.

    a. Compare ΔV_{ab}, ΔV_{cd}, and ΔV_{ef}. Are they all the same? If not, rank them in decreasing order. Explain your reasoning.

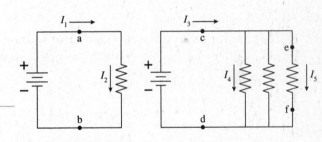

    b. Rank in order, from largest to smallest, the five currents I_1 to I_5.

    Order:

    Explanation:

Exercises 22–28: Assume that all wires are ideal (zero resistance) and that all batteries are ideal (constant potential difference).

22. Initially bulbs A and B are glowing. Then the switch is closed. What happens to each bulb? Does it get brighter, stay the same, get dimmer, or go out? Explain your reasoning.

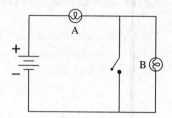

23. a. Bulbs A, B, and C are identical. Rank in order, from most to least, the brightnesses of the three bulbs.

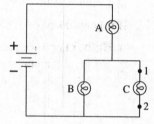

   Order:

   Explanation:

   b. Suppose a wire is connected between points 1 and 2. What happens to each bulb? Does it get brighter, stay the same, get dimmer, or go out? Explain.

24. a. Consider the points a and b. Is the potential difference $\Delta V_{ab} = 0$? If so, why? If not, which point is more positive?

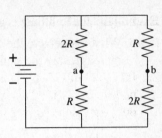

b. If a wire is connected between points a and b, does a current flow through it? If so, in which direction— to the right or to the left? Explain.

25. Bulbs A and B are identical. Initially both are glowing.

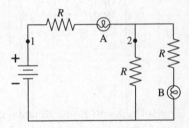

a. Bulb A is removed from its socket. What happens to bulb B? Does it get brighter, stay the same, get dimmer, or go out? Explain.

b. Bulb A is replaced. Bulb B is then removed from its socket. What happens to bulb A? Does it get brighter, stay the same, get dimmer, or go out? Explain.

c. The circuit is restored to its initial condition. A wire is then connected between points 1 and 2. What happens to the brightness of each bulb?

26. Initially the lightbulb is glowing. It is then removed from its socket.

    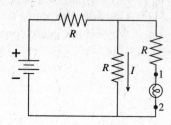

    a. What happens to the current *I* when the bulb is removed? Does it increase, stay the same, or decrease? Explain.

    b. What happens to the potential difference ΔV_{12} between points 1 and 2? Does it increase, stay the same, decrease, or become zero? Explain.

27. Bulbs A and B are identical and initially both are glowing. Then the switch is closed. What happens to each bulb? Does its brightness increase, stay the same, decrease, or go out? Explain.

    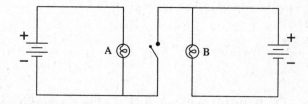

28. Bulbs A and B are identical and initially both are glowing. Then the switch is closed. What happens to each bulb? Does its brightness increase, stay the same, decrease, or go out? Explain.

    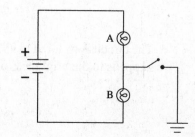

32.9 *RC* Circuits

29. The graph shows the voltage as a function of time on a capacitor as it is discharged (separately) through three different resistors. Rank in order, from largest to smallest, the values of the resistances R_1 to R_3.

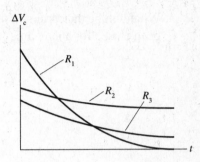

    Order:

    Explanation:

30. The capacitors in each circuit are discharged when the switch closes at $t = 0$ s. Rank in order, from largest to smallest, the time constants τ_1 to τ_5 with which each circuit will discharge.

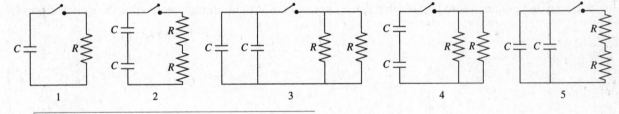

    Order:

    Explanation:

31. The charge on the capacitor is zero when the switch closes at $t = 0$ s.

    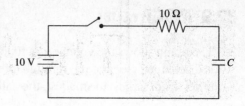

    a. What will be the current in the circuit after the switch has been closed for a long time? Explain.

    b. Immediately after the switch closes, before the capacitor has had time to charge, the potential difference across the capacitor is zero. What must be the potential difference across the resistor in order to satisfy Kirchhoff's loop law? Explain.

    c. Based on your answer to part b, what is the current in the circuit immediately after the switch closes?

    d. Sketch a graph of current versus time, starting from just before $t = 0$ s and continuing until the switch has been closed a long time. There are no numerical values for the horizontal axis, so you should think about the *shape* of the graph.

    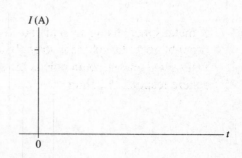

33 The Magnetic Field

33.1 Magnetism

1. A lightweight glass sphere hangs by a thread. The north pole of a bar magnet is brought near the sphere.

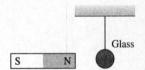
Glass

a. Suppose the sphere is electrically neutral. How does it respond?
   i.   It is strongly attracted to the magnet.
   ii. It is weakly attracted to the magnet.
   iii. It does not respond.
   iv. It is weakly repelled by the magnet.
   v.   It is strongly repelled by the magnet.

   Explain your choice.

b. How does the sphere respond if it is positively charged? Explain.

2. A metal sphere hangs by a thread. When the north pole of a bar magnet is brought near, the sphere is strongly attracted to the magnet. Then the magnet is reversed and its south pole is brought near the sphere. How does the sphere respond? Explain.

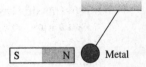
Metal

3. The compass needle below is free to rotate in the plane of the page. Either a bar magnet or a charged rod is brought toward the *center* of the compass. Does the compass rotate? If so, in which direction? If not, why not?

a.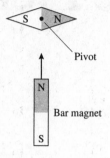

Pivot

Bar magnet

b.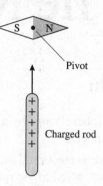

Pivot

Charged rod

4. You have two electrically neutral metal cylinders that exert strong attractive forces on each other. You have no other metal objects. Can you determine if *both* of the cylinders are magnets, or if one is a magnet and the other just a piece of iron? If so, how? If not, why not?

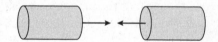

5. Can you think of any kind of object that is repelled by *both* ends of a bar magnet? If so, what? If not, what prevents this from happening?

33.2 The Discovery of the Magnetic Field

6. A neutral copper rod, a polarized insulator, and a bar magnet are arranged around a current-carrying wire as shown. For each, will it stay where it is? Move toward or away from the wire? Rotate clockwise or counterclockwise? Explain.

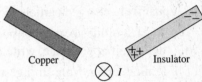
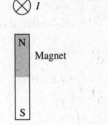

a. Neutral copper rod

b. Polarized insulator

c. Bar magnet

7. For each of the current-carrying wires shown, draw a compass needle in its equilibrium position at the positions of the dots. Label the poles of the compass needle.

a.

b.

8. The figure shows a wire directed into the page and a nearby compass needle. Is the wire's current going into the page or coming out of the page? Explain.

9. A compass is placed at 12 different positions and its orientation is recorded. Use this information to draw the magnetic *field lines* in this region of space. Draw the field lines on the figure.

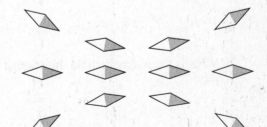

33.3 The Source of the Magnetic Field: Moving Charges

10. A positively charged particle moves toward the bottom of the page.

    a. At each of the six number points, show the direction of the magnetic field or, if appropriate, write $\vec{B} = \vec{0}$.

    b. Rank in order, from strongest to weakest, the magnetic field strengths B_1 to B_6 at these points.

    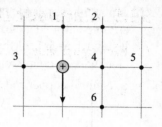

    Order:

    Explanation:

11. The negative charge is moving out of the page, coming toward you. Draw the magnetic field lines in the plane of the page.

12. Two charges are moving as shown. At this instant of time, the net magnetic field at point 2 is $\vec{B}_2 = \vec{0}$.

    a. Is the unlabeled moving charge positive or negative? Explain.

    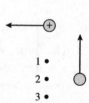

    b. What is the magnetic field direction at point 1? Explain.

    c. What is the magnetic field direction at point 3?

© 2008 by Pearson Education, Inc., publishing as Pearson Addison-Wesley.

33.4 The Magnetic Field of a Current

33.5 Magnetic Dipoles

13. Each figure shows a current-carrying wire. Draw the magnetic field diagram:

a.

b.

The wire is perpendicular to the page. Draw magnetic field *lines*, then show the magnetic field *vectors* at a few points around the wire.

The wire is in the plane of the page. Show the magnetic field above and below the wire.

14. This current-carrying wire is in the plane of the page. Draw the magnetic field on both sides of the wire.

15. Use an arrow to show the current direction in this wire.

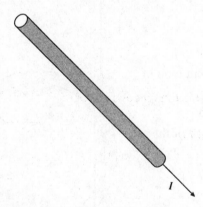

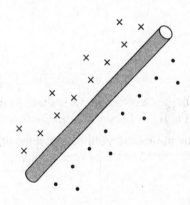

16. Each figure below shows two long straight wires carrying equal currents into and out of the page. At each of the dots, use a **black** pen or pencil to show and label the magnetic fields $\vec{B}_1$ and $\vec{B}_2$ of each wire. Then use a **red** pen or pencil to show the net magnetic field.

a.

Wire 1

⊗

•   •

⊙

Wire 2

b.

Wire 1

⊙

•   •

⊙

Wire 2

17. A long straight wire, perpendicular to the page, passes through a uniform magnetic field. The *net* magnetic field at point 3 is zero.

    a. On the figure, show the direction of the current in the wire.

    b. Points 1 and 2 are the same distance from the wire as point 3, and point 4 is twice as distant. Construct vector diagrams at points 1, 2, and 4 to determine the net magnetic field at each point.

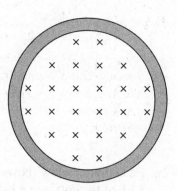

18. A long straight wire passes above one edge of a current loop. Both are perpendicular to the page. $\vec{B}_1 = \vec{0}$ at point 1.

    a. On the figure, show the direction of the current in the loop.

    b. Use a vector diagram to determine the net magnetic field at point 2.

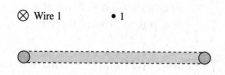

19. The figure shows the magnetic field seen when facing a current loop in the plane of the page.

    a. On the figure, show the direction of the current in the loop.

    b. Is the north pole of this loop at the upper surface of the page or the lower surface of the page? Explain.

20. The current loop exerts a repulsive force on the bar magnet. On the figure, show the direction of the current in the loop. Explain.

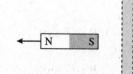

33.6 Ampère's Law and Solenoids

21. What is the total current through the area bounded by the closed curve?

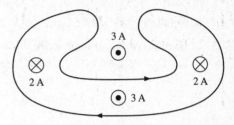

22. The total current through the area bounded by the closed curve is 2 A. What are the size and direction of I_3?

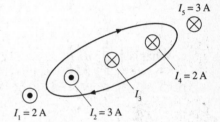

23. The magnetic field above the dotted line is $\vec{B} = (2\text{ T, right})$. Below the dotted line the field is $\vec{B} = (2\text{ T, left})$. Each closed loop is 1 m × 1 m. Let's evaluate the line integral of $\vec{B}$ around each of these closed loops by breaking the integration into four steps. We'll go around the loop in a *clockwise* direction. Pay careful attention to signs.

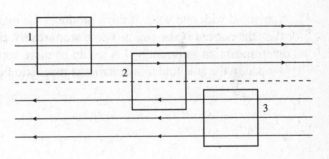

	Loop 1	Loop 2	Loop 3
$\int \vec{B} \cdot d\vec{s}$ along left edge			
$\int \vec{B} \cdot d\vec{s}$ along top			
$\int \vec{B} \cdot d\vec{s}$ along right edge			
$\int \vec{B} \cdot d\vec{s}$ along bottom			

The line integral *around* the loop is simply the sum of these four separate integrals:

$\oint \vec{B} \cdot d\vec{s}$ around the loop

24. The strength of a circular magnetic field decreases with increasing radius as shown.

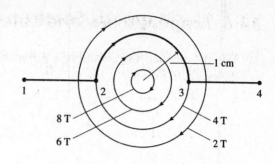

a. What is $\int_1^2 \vec{B} \cdot d\vec{s}$? _____

Explain or show your work.

b. What is $\int_2^3 \vec{B} \cdot d\vec{s}$? _____   Explain or show your work.

c. What is $\int_3^4 \vec{B} \cdot d\vec{s}$? _____   Explain or show your work.

d. Combining your answers to parts a to c, what is $\int_1^4 \vec{B} \cdot d\vec{s}$? _____

25. A solenoid with one layer of turns produces the magnetic field strength you need for an experiment when the current in the coil is 3 A. Unfortunately, this amount of current overheats the coil. You've determined that a current of 1 A would be more appropriate. How many additional layers of turns must you add to the solenoid to maintain the magnetic field strength? Explain.

26. Rank in order, from largest to smallest, the magnetic fields B_1 to B_3 produced by these three solenoids.

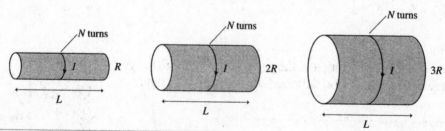

Order:

Explanation:

33.7 The Magnetic Force on a Moving Charge

27. For each of the following, draw the magnetic force vector on the charge or, if appropriate, write "$\vec{F}$ into page," "$\vec{F}$ out of page," or "$\vec{F} = \vec{0}$."

a.

b.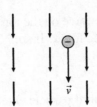

c.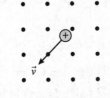

d.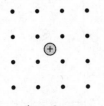

$\vec{v}$ out of page

e.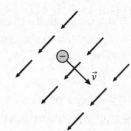

f.

28. For each of the following, determine the sign of the charge (+ or −).

a.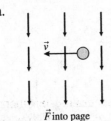

$\vec{F}$ into page

$q =$ _____

b.

$\vec{v}$ into page

$q =$ _____

c.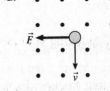

$q =$ _____

d.

$q =$ _____

29. The magnetic field is constant magnitude inside the dashed lines and zero outside. Sketch and label the trajectory of the charge for

a. A weak field.

b. A strong field.

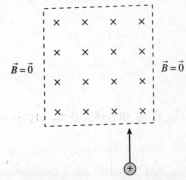

30. A positive ion, initially traveling into the page, is shot through the gap in a horseshoe magnet. Is the ion deflected up, down, left, or right? Explain.

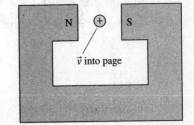

31. A positive ion is shot between the plates of a parallel-plate capacitor.

    a. In what direction is the electric force on the ion?

    _____

    b. Could a magnetic field exert a magnetic force on the ion that is opposite in direction to the electric force? If so, show the magnetic field on the figure.

    _____

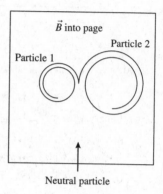

32. In a high-energy physics experiment, a neutral particle enters a bubble chamber in which a magnetic field points into the page. The neutral particle undergoes a collision inside the bubble chamber, creating two charged particles. The subsequent trajectories of the charged particles are shown.

    a. What is the sign (+ or −) of particle 1? _____

    What is the sign (+ or −) of particle 2? _____

    b. Which charged particle leaves the collision with a larger momentum? Explain. (Assume that $|q| = e$ for both particles.)

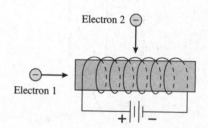

33. A solenoid is wound as shown and attached to a battery. Two electrons are fired into the solenoid, one from the end and one through a very small hole in the side.

    a. In what direction does the magnetic field inside the solenoid point? Show it on the figure.

    b. Is electron 1 deflected as it moves through the solenoid? If so, in which direction? If not, why not?

    _____

    c. Is electron 2 deflected as it moves through the solenoid? If so, in which direction? If not, why not?

    _____

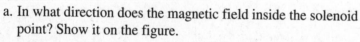

34. Two protons are traveling in the directions shown.

    a. Draw and label the electric force on each proton due to the other proton.

    b. Draw and label the magnetic force on each proton due to the other proton. Explain how you determined the directions.

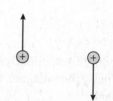

33.8 Magnetic Forces on Current-Carrying Wires

33.9 Forces and Torques on Current Loops

35. Three current-carrying wires are perpendicular to the page. Construct a force vector diagram on the figure to find the net force on the upper wire due to the two lower wires.

36. Three current-carrying wires are perpendicular to the page.

    a. Construct a force vector diagram on each wire to determine the direction of the net force on each wire.

    b. Can three *charges* be placed in a triangular pattern so that their force diagram looks like this? If so, draw it below. If not, why not?

37. A current-carrying wire passes in front of a solenoid that is wound as shown. The wire experiences an upward force. Use arrows to show the direction in which the current enters and leaves the solenoid. Explain your choice.

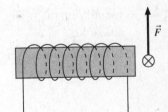

38. A current loop is placed between two bar magnets. Does the loop move to the right, move to the left, rotate clockwise, rotate counterclockwise, some combination of these, or none of these? Explain.

39. A square current loop is placed in a magnetic field as shown.

    a. Does the loop undergo a displacement? If so, is it up, down, left, or right? If not, why not?

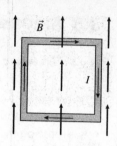

    b. Does the loop rotate? If so, which edge rotates out of the page and which edge into the page? If not, why not?

40. The south pole of a bar magnet is brought toward the current loop. Does the bar magnet attract the loop, repel the loop, or have no effect on the loop? Explain.

33.10 Magnetic Properties of Matter

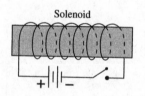

41. A solenoid, wound as shown, is placed next to an unmagnetized piece of iron. Then the switch is closed.

    a. Identify on the figure the north and south poles of the solenoid.

    b. What is the direction of the solenoid's magnetic field as it passes through the iron?

    c. What is the direction of the induced magnetic dipole in the iron?

    d. Identify on the figure the north and south poles of the induced magnetic dipole in the iron.

    e. When the switch is closed, does the iron move left or right? Does it rotate? Explain.

34 Electromagnetic Induction

34.1 Induced Currents

34.2 Motional emf

1. The figures below show one or more metal wires sliding on fixed metal rails in a magnetic field. For each, determine if the induced current flows clockwise, flows counterclockwise, or is zero. Show your answer by drawing it.

a.
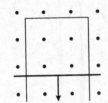

b.

c.
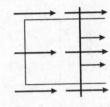

d.
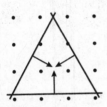

e.
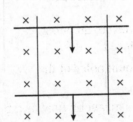

f.
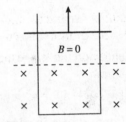

2. A loop of copper wire is being pulled from between two magnetic poles.

   a. Show on the figure the current induced in the loop. Explain your reasoning.

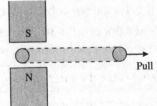

   b. Does either side of the loop experience a magnetic force? If so, draw and label a vector arrow or arrows on the figure to show any forces.

3. A vertical, rectangular loop of copper wire is half in and half out of a horizontal magnetic field. (The field is zero beneath the dotted line.) The loop is released and starts to fall.

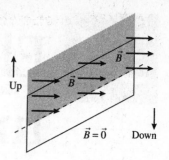

a. Add arrows to the figure to show the direction of the induced current in the loop.

b. Is there a net magnetic force on the loop? If so, in which direction? Explain.

4. Two very thin sheets of copper are pulled through a magnetic field. Do eddy currents flow in the sheet? If so, show them on the figures, with arrows to indicate the direction of flow. If not, why not?

a.

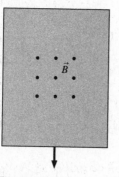

b.

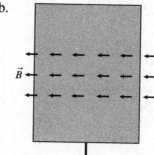

5. The figure shows an edge view of a copper sheet being pulled between two magnetic poles.

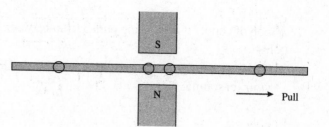

a. Add a dot or an × to each of the circles to indicate the direction in which eddy currents are flowing in and out of the page.

b. Do the currents you labeled in part a experience magnetic forces? If so, add force vectors to the figure to show the directions. If not, why not?

c. Is there a net magnetic force on the copper sheet? If so, in which direction?

34.3 Magnetic Flux

6. The figure shows five loops in a magnetic field. The numbers indicate the lengths of the sides and the strength of the field. Rank in order, from largest to smallest, the magnetic fluxes Φ_1 to Φ_5. Some may be equal.

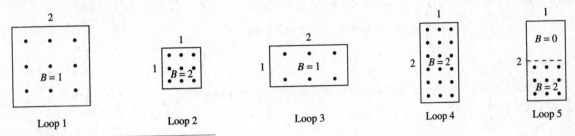

Order:

Explanation:

7. The figure shows four circular loops that are perpendicular to the page. The radius of loops 3 and 4 is twice that of loops 1 and 2. The magnetic field is the same for each. Rank in order, from largest to smallest, the magnetic fluxes Φ_1 to Φ_4. Some may be equal.

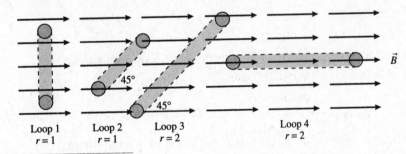

Order:

Explanation:

8. A circular loop rotates at constant speed about an axle through the center of the loop. The figure shows an edge view and defines the angle ϕ, which increases from 0° to 360° as the loop rotates.

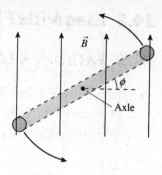

   a. At what angle or angles is the magnetic flux a maximum?

   b. At what angle or angles is the magnetic flux a minimum?

   c. At what angle or angles is the magnetic flux *changing* most rapidly? Explain your choice.

9. A magnetic field is perpendicular to a loop. The graph shows how the magnetic field changes as a function of time, with positive values for B indicating a field into the page and negative values a field out of the page. Several points on the graph are labeled.

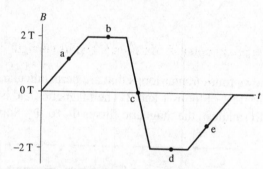

Field through loop

   a. At which lettered point or points is the flux through the loop a maximum?

   b. At which lettered point or points is the flux through the loop a minimum?

   c. At which point or points is the flux changing most rapidly?

   d. At which point or points is the flux changing least rapidly?

34.4 Lenz's Law

34.5 Faraday's Law

10. Does the loop of wire have a clockwise current, a counterclockwise current, or no current under the following circumstances? Explain.

   a. The magnetic field points into the page and its strength is increasing.

   b. The magnetic field points into the page and its strength is constant.

   c. The magnetic field points into the page and its strength is decreasing.

11. A loop of wire is perpendicular to a magnetic field. The magnetic field strength as a function of time is given by the top graph. Draw a graph of the current in the loop as a function of time. Let a positive current represent a current that comes out of the top and enters the bottom. There are no numbers for the vertical axis, but your graph should have the correct shape and proportions.

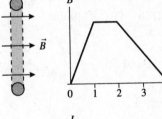

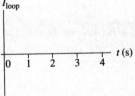

12. A loop of wire is horizontal. A bar magnet is pushed toward the
    loop from below, along the axis of the loop.

    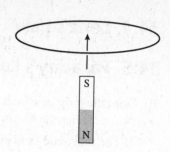

    a. What is the current direction in the loop? Explain.

    b. Is there a magnetic force on the loop? If so, in which direction? Explain.
       Hint: A current loop is a magnetic dipole.

13. A bar magnet is dropped, south pole down, through the center of a
    loop of wire. The center of the magnet passes the plane of the loop
    at time t_c.

    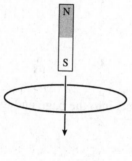

    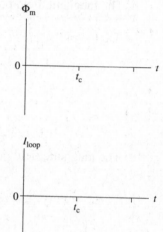

    a. Sketch a graph of the magnetic flux through
       the loop as a function of time.

    b. Sketch a graph of the current in the loop as a
       function of time. Let a clockwise current be a
       positive number and a counterclockwise
       current be a negative number.

14. a. Just after the switch on the left coil is closed,
       does current flow right to left or left to right
       through the current meter of the right coil?
       Or is the current zero? Explain.

    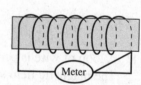
    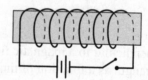

    b. Long after the switch on the left coil is closed, does current flow right to left or left to right
       through the current meter of the right coil? Or is the current zero? Explain.

15. A solenoid is perpendicular to the page, and its field strength is increasing. Three circular wire loops of equal radii are shown. Rank in order, from largest to smallest, the size of the induced emf in the three rings.

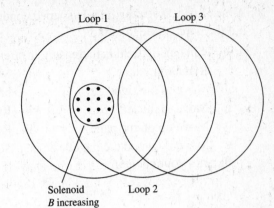

Loop 1   Loop 3

Solenoid
B increasing   Loop 2

Order:

Explanation:

16. A conducting loop around a magnetic field contains two lightbulbs, A and B. The wires connecting the bulbs are ideal, with no resistance. The magnetic field is increasing rapidly.

    a. Do the bulbs glow? Why or why not?

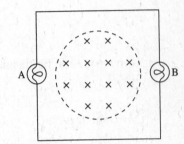

    b. If they glow, which bulb is brighter? Or are they equally bright? Explain.

17. A conducting loop around a magnetic field contains three lightbulbs, A, B, and C. The wires connecting the bulbs are ideal, with no resistance. The magnetic field is increasing rapidly. Rank in order, from brightest to least bright, the brightness of the three bulbs.

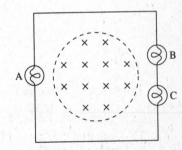

Order:

Explanation:

18. A metal wire is resting on a U-shaped conducting rail. The rail is fixed in position, but the wire is free to move.

    a. If the magnetic field is increasing in strength, does the wire:

    i.   Remain in place?
    ii. Move to the right?
    iii. Move to the left?
    iv. Move up on the page?
    v.   Move down on the page?

    vi. Move out of the plane of the page, breaking contact with the rail?
    vii. Rotate clockwise?
    viii. Rotate counterclockwise?
    ix. Some combination of these? If so, which?

    Explain your choice.

    b. If the magnetic field is decreasing in strength, which of the above happens? Explain.

34.6 Induced Fields

19. Consider these two situations:

    a. Draw the induced electric field.

    $\vec{B}$-field rapidly increasing

    b. Draw the induced electric field.

    $\vec{B}$-field rapidly decreasing

34.7 Induced Current: Three Applications

No exercises.

34.8 Inductors

20. The figure shows the current through an inductor. A positive current is defined as a current going from top to bottom. At the time corresponding to each of the labeled points, does the potential across the inductor (going from top to bottom) increase, decrease, or stay the same?

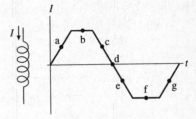

   a. _____   e. _____

   b. _____   f. _____

   c. _____   g. _____

   d. _____

21. Rank in order, from most positive to most negative, the inductor's potential difference $(\Delta V_L)_a$, $(\Delta V_L)_b$, ..., $(\Delta V_L)_f$, at the six labeled points. ΔV_L is the change in going from the top of the inductor to the bottom. Some may be equal. Note that $0\,V > -2\,V$.

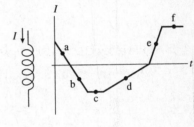

   Order:

   Explanation:

22. The figure shows the current through an inductor. Draw a graph showing the potential difference ΔV_L across the inductor. There are no numbers, but your graph should have the correct shape and proportions.

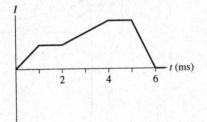
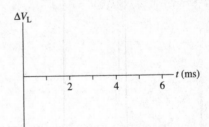

23. The figure shows the potential difference across an inductor. There is no current at $t = 0$. Draw a graph of the current through the inductor as a function of time. There are no numbers, but your graph should have the correct shape and proportions.

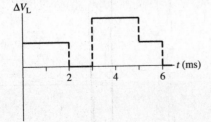
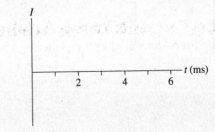

34.9 *LC* Circuits

24. An *LC* circuit oscillates at a frequency of 2000 Hz. What will the frequency be if the inductance is quadrupled?

25. The capacitor in an *LC* circuit has maximum charge at $t = 1$ μs. The current through the inductor next reaches a maximum at $t = 3$ μs.

    a. When will the inductor current reach a maximum in the opposite direction?

    b. What is the circuit's period of oscillation?

34.10 *LR* Circuits

26. Three *LR* circuits are made with the same resistor but different inductors. The figure shows the inductor current as a function of time. Rank in order, from largest to smallest, the three inductances L_1, L_2, and L_3.

    Order:

    Explanation:

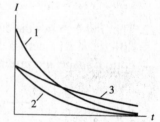

27. a. What is the battery current immediately after the switch closes? Explain.

    b. What is the battery current after the switch has been closed a long time? Explain.

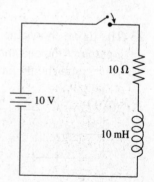

35 Electromagnetic Fields and Waves

35.1 *E* or *B*? It Depends on Your Perspective

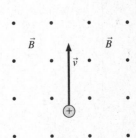

1. In frame S, a positive charge moves through the magnetic field shown.
   a. Draw a vector on the charge to show the magnetic force in S.
   b. What are the speed *V* and direction of a reference frame S′ in which there is no magnetic force? Explain.

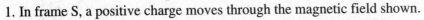

   c. What are the type and direction of any fields in S′ that could cause the observed force on the charge?

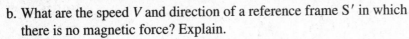

2. Sharon drives her rocket through a magnetic field, traveling to the right at a speed of 1000 m/s as measured by Bill. As she passes Bill, she shoots a positive charge backward at a speed of 1000 m/s relative to her.

   a. According to Bill, what kind of force or forces act on the charge? In which directions? Explain.

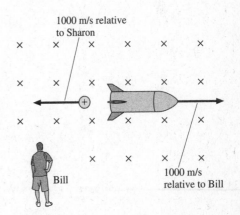

   b. According to Sharon, what kind of force or forces act on the charge? In which directions? Draw the forces on the charge.

3. In frame S, a positive charge moves to the right at speed v. Frame S′ travels to the right at speed $V = v$ relative to S. Frame S″ travels to the right at speed $V = 2v$ relative to S. The figure below shows the charge three times, once in each reference frame.

   a. For each:

   • Draw and label a velocity vector on the charge showing its motion in that frame.

   • Draw and label the electric and magnetic field vectors due to the charge at the points above and below the charge. Use the notation of circled $\times$ and $\bullet$ to show fields into or out of the page.

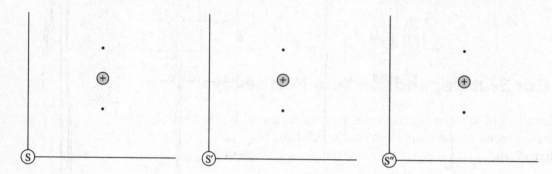

   b. Does it make sense to talk about "the" magnetic field? Why or why not?

35.2 The Field Laws Thus Far

35.3 The Displacement Current

4. If you curl the fingers of your right hand as shown, is the electric flux positive or negative?

a.

b.

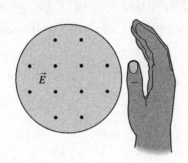

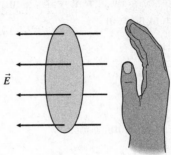

Sign of Φ_e _____   Sign of Φ_e _____

5. If you curl the fingers of your right hand as shown, is the emf positive or negative?

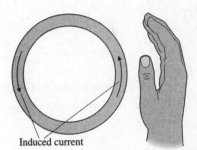

Induced current

6. What is the current through surface S?

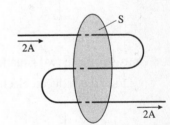

7. The capacitor in this circuit was initially charged, then the switch was closed. At this instant of time, the potential difference across the resistor is $\Delta V_R = 4$ V.

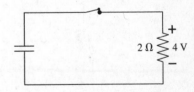

a. At this instant of time, what is the current through the resistor?

b. At this instant of time, what is the current through the space between the capacitor plates?

c. At this instant of time, what is the displacement current through the space between the capacitor plates?

d. Is the displacement current really a current? If so, what are the moving charges? If not, what is the displacement current?

8. Consider these two situations:

a.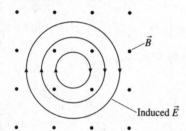

Is the magnetic field strength increasing, decreasing, or not changing? Explain.

b.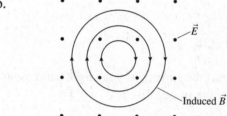

Is the electric field strength increasing, decreasing, or not changing? Explain.

9. Consider these two situations:

a. Draw the induced electric field.

b. Draw the induced magnetic field.

×	×	×	×
×	×	×	×
×	×	×	×
×	×	×	×

$\vec{B}$-field rapidly increasing

×	×	×	×
×	×	×	×
×	×	×	×
×	×	×	×

$\vec{E}$-field rapidly increasing

35.4 Maxwell's Equations

35.5 Electromagnetic Waves

35.6 Properties of Electromagnetic Waves

10. This is an electromagnetic plane wave traveling into the page. Draw the magnetic field vectors $\vec{B}$ at the dots.

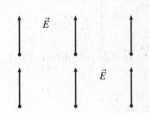

11. This is an electromagnetic wave at one instant of time.

    a. Draw the velocity vector $\vec{v}_{em}$.

    b. Draw $\vec{E}$, $\vec{B}$, and $\vec{v}_{em}$ a half cycle later.

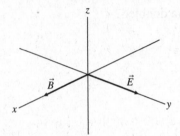

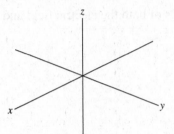

12. Do the following represent possible electromagnetic waves? If not, why not?

    a.

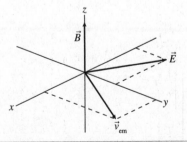

    b.

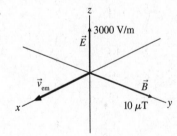

    c.

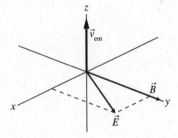

    d.

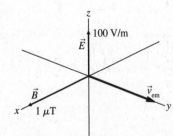

13. The intensity of an electromagnetic wave is 10 W/m^2. What will be the intensity if:

   a. The amplitude of the electric field is doubled?

   b. The amplitude of the magnetic field is doubled?

   c. The amplitudes of both the electric field and the magnetic field are doubled?

   d. The frequency is doubled?

35.7 Polarization

14. A polarized electromagnetic wave passes through a polarizing filter. Draw the electric field of the wave after it has passed through the filter.

a.

b.

15. A polarized electromagnetic wave passes through a series of polarizing filters. Draw the electric field of the wave after it has passed through each filter.

a.

b.

16. The intensity of a polarized electromagnetic wave is 10 W/m^2. What will be the intensity of the wave after it passes through a polarizing filter whose axis makes the following angle with the plane of polarization?

$\theta = 0°$ _____ $\theta = 60°$ _____

$\theta = 30°$ _____ $\theta = 90°$ _____

$\theta = 45°$ _____

36 AC Circuits

36.1 AC Sources and Phasors

1. The figure shows emf phasors A, B, and C.

   a. What is the instantaneous value of the emf?

   A _____ B _____ C _____

   b. At this instant, is the magnitude of the emf increasing, decreasing, or holding constant?

   A _____ B _____ C _____

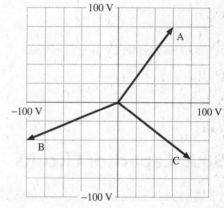

2. Draw a phasor diagram for the following emfs.

   a. $(100\text{ V})\cos\omega t$ at $\omega t = 240°$ b. $(400\text{ V})\cos\omega t$ at $t = \frac{1}{3}T$ c. $(200\text{ V})\cos\omega t$ at $t = 0$

3. The current phasor is shown for a 10 Ω resistor.

   a. What is the instantaneous resistor voltage v_R?

   b. What is the peak resistor voltage V_R?

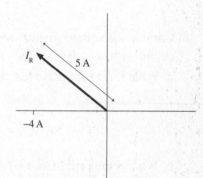

4. The peak current through a resistor is 4.0 A. What is the peak current if:

   a. The resistance R is doubled?

   b. The peak emf $\mathcal{E}_0$ is doubled?

   c. The frequency ω is doubled?

36.2 Capacitor Circuits

5. The peak current through a capacitor is 4.0 A. What is the peak current if:

   a. The peak emf $\mathcal{E}_0$ is doubled?

   b. The capacitance C is doubled?

   c. The frequency ω is doubled?

6. Current and voltage graphs are shown for a capacitor circuit with $\omega = 1000$ rad/s.

   a. What is the capacitive reactance X_C?

   b. What is the capacitance C?

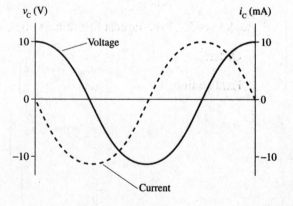

7. A 13 μF capacitor is connected to a 5.5 V/250 Hz oscillator. What is the instantaneous capacitor current i_C when $\mathcal{E} = -5.5$ V?

8. Consider these three circuits.

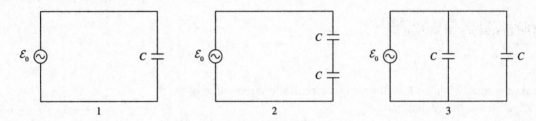

Rank in order, from largest to smallest, the peak currents $(I_C)_1$ to $(I_C)_3$ provided by the emf.

Order:

Explanation:

9. Consider these four circuits.

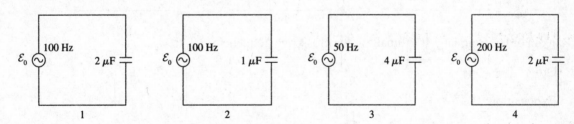

Rank in order, from largest to smallest, the capacitive reactances $(X_C)_1$ to $(X_C)_4$.

Order:

Explanation:

36.3 *RC* Filter Circuits

10. A low-pass *RC* filter has a crossover frequency f_c = 200 Hz. What is f_c if:

a. The resistance *R* is halved?

b. The capacitance *C* is halved?

c. The peak emf $\mathcal{E}_0$ is halved?

11. What new resistor value *R* will give this circuit the same value of ω_c if the capacitor value is changed to:

a. *C* = 1 μF   *R* = _____

b. *C* = 4 μF   *R* = _____

c. *C* = 20 μF   *R* = _____

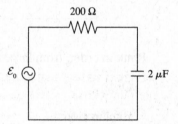

12. Consider these three circuits.

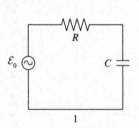   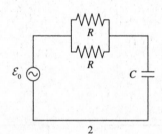   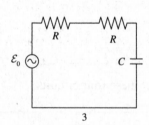

Rank in order, from largest to smallest, the crossover frequencies ω_{c1} to ω_{c3}.

Order:

Explanation:

13. The text claims that $V_R = V_C = \mathcal{E}_0/\sqrt{2}$ at $\omega = \omega_c$. If this is true, then $V_R + V_C > \mathcal{E}_0$. Is it possible for their sum to be larger than $\mathcal{E}_0$? Explain.

36.4 Inductor Circuits

14. The peak current passing through an inductor is 4.0 A. What is the peak current if:

    a. The peak emf $\mathcal{E}_0$ is doubled?

    b. The inductance L is doubled?

    c. The frequency ω is doubled?

15. Current and voltage graphs are shown for an inductor circuit with $\omega = 1000$ rad/s.

    a. What is the inductive reactance X_L?

    b. What is the inductance L?

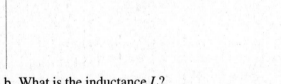

16. Consider these four circuits.

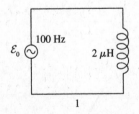 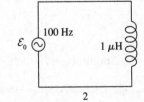 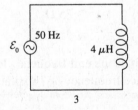 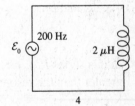

Rank in order, from largest to smallest, the inductive reactances $(X_L)_1$ to $(X_L)_4$.

Order:

Explanation:

36.5 The Series *RLC* Circuit

17. The resonance frequency of a series *RLC* circuit is 1000 Hz. What is the resonance frequency if:

    a. The resistance *R* is doubled?

    b. The inductance *L* is doubled?

    c. The capacitance *C* is doubled?

    d. The peak emf $\mathcal{E}_0$ is doubled?

    e. The frequency ω is doubled?

18. For these combinations of resistance and reactance, is a series *RLC* circuit in resonance (Yes or No)? Does the current lead the emf, lag the emf, or is it in phase with the emf?

R	X_L	X_C	Resonance?	Current?
100 Ω	100 Ω	50 Ω		
100 Ω	50 Ω	100 Ω		
100 Ω	75 Ω	75 Ω		

19. In this series *RLC* circuit, is the emf frequency less than, equal to, or greater than the resonance frequency ω_0? Explain.

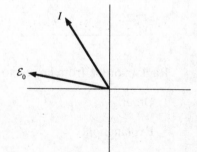

© 2008 by Pearson Education, Inc., publishing as Pearson Addison-Wesley.

20. The resonance frequency of a series *RLC* circuit is greater than the emf frequency. Does the current lead or lag the emf? Explain.

21. Consider these four circuits. They all have the same resonance frequency ω_0.

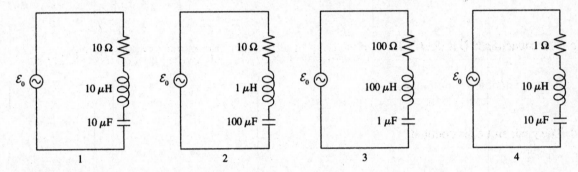

Rank in order, from largest to smallest, the maximum currents $(I_{max})_1$ to $(I_{max})_4$.

Order:

Explanation:

22. The current in a series *RLC* circuit lags the emf by 20°. You cannot change the emf. What two different things could you do to the circuit that would increase the power delivered to the circuit by the emf?

36.6 Power in AC Circuits

23. An average power dissipated by a resistor is 4.0 W. What is P_{avg} if:

   a. The resistance R is doubled?

   b. The peak emf $\mathcal{E}_0$ is doubled?

   c. Both are doubled simultaneously?

24. Consider these three circuits.

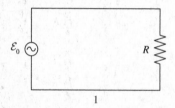
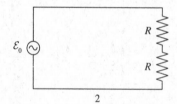
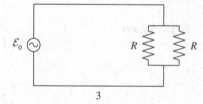

Rank in order, from largest to smallest, the average powers P_1 to P_3 delivered by the three emfs.

Order:

Explanation:

37 Relativity

37.1 Relativity: What's It All About?

37.2 Galilean Relativity

1. In which reference frame, S or S′, does the ball move faster?

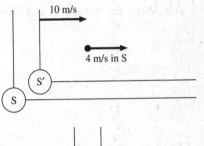

2. Frame S′ moves relative to frame S as shown.

   a. A ball is at rest in frame S′. What are the speed and direction of the ball in frame S?

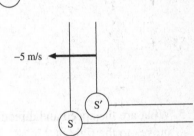

   b. A ball is at rest in frame S. What are the speed and direction of the ball in frame S′?

3. Frame S′ moves parallel to the x-axis of frame S.

   a. Is there a value of v for which the ball is at rest in S′? If so, what is v? If not, why not?

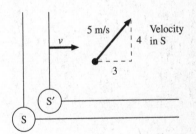

   b. Is there a value of v for which the ball has a minimum speed in S′? If so, what is v? If not, why not?

4. a. What are the speed and direction of each ball in a reference frame that moves with ball 1?

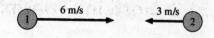

b. What are the speed and direction of each ball in a reference frame that moves with ball 2?

5. What are the speed and direction of each ball in a reference frame that moves to the right at 2 m/s?

37.3 Einstein's Principle of Relativity

6. A lighthouse beacon alerts ships to the danger of a rocky coastline.

   a. According to the lighthouse keeper, with what speed does the light leave the lighthouse?

   b. A boat is approaching the coastline at speed $0.5c$. According to the captain, with what speed is the light from the beacon approaching her boat?

7. As a rocket passes the earth at $0.75c$, it fires a laser perpendicular to its direction of travel.

   a. What is the speed of the laser beam relative to the rocket?

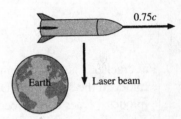

   b. What is the speed of the laser beam relative to the earth?

8. Teenagers Sam and Tom are playing chicken in their rockets. As seen from the earth, each is traveling at $0.95c$ as he approaches the other. Sam fires a laser beam toward Tom.

   a. What is the speed of the laser beam relative to Sam?

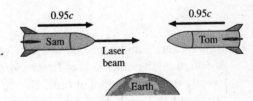

   b. What is the speed of the laser beam relative to Tom?

37.4 Events and Measurements

9. It is a bitter cold day at the South Pole, so cold that the speed of sound is only 300 m/s. The speed of light, as always, is 300 m/μs. A firecracker explodes 600 m away from you.

  a. How long after the explosion until you see the flash of light? _____

  b. How long after the explosion until you hear the sound? _____

  c. Suppose you see the flash at $t = 2.000002$ s. At what time was the explosion? _____

  d. What are the spacetime coordinates for the event "firecracker explodes"? Assume that you are at the origin and that the explosion takes place at a position on the positive x-axis.

10. Firecracker 1 is 300 m from you. Firecracker 2 is 600 m from you in the same direction. You see both explode at the same time. Define event 1 to be "firecracker 1 explodes" and event 2 to be "firecracker 2 explodes." Does event 1 occur before, after, or at the same time as event 2? Explain.

11. Firecrackers 1 and 2 are 600 m apart. You are standing exactly halfway between them. Your lab partner is 300 m on the other side of firecracker 1. You see two flashes of light, from the two explosions, at exactly the same instant of time. Define event 1 to be "firecracker 1 explodes" and event 2 to be "firecracker 2 explodes." According to your lab partner, based on measurements he or she makes, does event 1 occur before, after, or at the same time as event 2? Explain.

12. Two trees are 600 m apart. You are standing exactly halfway between them and your lab partner is at the base of tree 1. Lightning strikes both trees.

   a. Your lab partner, based on measurements he or she makes, determines that the two lightning strikes were simultaneous. What did you see? Did you see the lightning hit tree 1 first, hit tree 2 first, or hit them both at the same instant of time? Explain.

   b. Lightning strikes again. This time your lab partner sees both flashes of light at the same instant of time. What did you see? Did you see the lightning hit tree 1 first, hit tree 2 first, or hit them both at the same instant of time? Explain.

   c. In the scenario of part b, were the lightning strikes simultaneous? Explain.

13. You are at the origin of a coordinate system containing clocks, but you're not sure if the clocks have been synchronized. The clocks have reflective faces, allowing you to read them by shining light on them. You flash a bright light at the origin at the instant your clock reads $t = 2.000000$ s.

   a. At what time will you see the reflection of the light from a clock at $x = 3000$ m?

   b. When you see the clock at $x = 3000$ m, it reads 2.000020 s. Is the clock synchronized with your clock at the origin? Explain.

37.5 The Relativity of Simultaneity

14. Two supernovas, labeled L and R, occur on opposite sides of a galaxy, at equal distances from the center. The supernovas are seen at the same instant on a planet at rest in the center of the galaxy. A spaceship is entering the galaxy from the left at a speed of 0.999c relative to the galaxy.

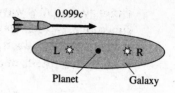

a. According to astronomers on the planet, were the two explosions simultaneous? Explain why.

b. Which supernova, L or R, does the spaceship crew *see* first? _____

c. Did the supernova that was *seen* first necessarily *happen* first in the rocket's frame? Explain.

d. Is "two light flashes reach the planet at the same instant" an event? To help you decide, could you arrange for something to happen only if two light flashes from opposite directions arrive at the same time? Explain.

If you answered Yes to part d, then the crew on the spaceship will also determine, from their measurements, that the light flashes reach the planet at the same instant. (Experimenters in different reference frames may disagree about when and where an event occurs, but they all agree that it *does* occur.)

e. The figure below shows the supernovas in the spaceship's reference frame with the *assumption* that the supernovas are simultaneous. The second half of the figure is a short time after the explosions. Draw two circular wave fronts on the second half of the figure to show the light from each supernova. Neither wave front has yet reached the planet. Be sure to consider:
- The points on which the wave fronts are centered.
- The wave speeds of each wave front.

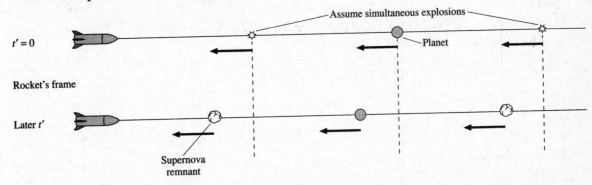

f. According to your diagram, are the two wave fronts going to reach the planet at the same instant of time? Why or why not?

g. Does your answer to part f conflict with your answer to part d? _____
If so, what different assumption could you make about the supernovas in the rocket's frame that would bring your wave-front diagram into agreement with your answer to part d?

h. So according to the spaceship crew, are the two supernovas simultaneous? If not, which happens first?

15. Peggy is standing at the center of her railroad car as it passes Ryan on the ground. Firecrackers attached to the ends of the car explode. A short time later, the flashes from the two explosions arrive at Peggy at the same time.

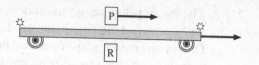

   a. Were the explosions simultaneous in Peggy's reference frame? If not, which exploded first? Explain.

   b. Were the explosions simultaneous in Ryan's reference frame? If not, which exploded first? Explain.

16. A rocket is traveling from left to right. At the instant it is halfway between two trees, lightning simultaneously (in the rocket's frame) hits both trees.

   a. Do the light flashes reach the rocket pilot simultaneously? If not, which reaches him first? Explain.

   b. A student was sitting on the ground halfway between the trees as the rocket passed overhead. According to the student, were the lightning strikes simultaneous? If not, which tree was hit first? Explain.

37.6 Time Dilation

17. Clocks C_1 and C_2 in frame S are synchronized. Clock C' moves at speed v relative to frame S. Clocks C' and C_1 read exactly the same as C' goes past. As C' passes C_2, is the time shown on C' earlier than, later than, or the same as the time shown on C_2? Explain.

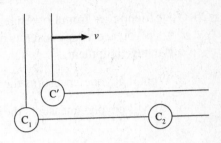

18. Your friend flies from Los Angeles to New York. She carries an accurate stopwatch with her to measure the flight time. You and your assistants on the ground also measure the flight time.

    a. Identify the two events associated with this measurement.

    b. Who, if anyone, measures the proper time? _____

    c. Who, if anyone, measures the shorter flight time? _____

19. You're passing a car on the highway. You want to know how much time is required to completely pass the car, from no overlap between the cars to no overlap between the cars. Call your car A and the car you are passing B.

    a. Specify two events that can be given spacetime coordinates. In describing the events, refer to cars A and B and to their front bumpers and rear bumpers.

    b. In either reference frame, is there *one* clock that is present at both events? _____

    c. Who, if anyone, measures the proper time between the events? _____

37.7 Length Contraction

20. Your friend flies from Los Angeles to New York. He determines the distance using the tried-and-true $d = vt$. You and your assistants on the ground also measure the distance, using meter sticks and surveying equipment.

    a. Who, if anyone, measures the proper length? _____

    b. Who, if anyone, measures the shorter distance? _____

21. Experimenters in B's reference frame measure $L_A = L_B$. Do experimenters in A's reference frame agree that A and B are the same length? If not, which do they find to be longer? Explain.

22. As a meter stick flies past you, you simultaneously measure the positions of both ends and determine that $L < 1$ m.

    a. To an experimenter in frame S', the meter stick's frame, did you make your two measurements simultaneously? If not, which end did you measure first? Explain.

       Hint: Review the reasoning about simultaneity that you used in Exercises 14–16.

    b. Can experimenters in frame S' give an explanation for why your measurement is < 1 m?

37.8 The Lorentz Transformations

23. A rocket travels at speed $0.5c$ relative to the earth.

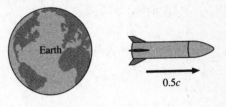

   a. The rocket shoots a bullet in the forward direction at speed $0.5c$ relative to the rocket. Is the bullet's speed relative to the earth less than, greater than, or equal to c?

   b. The rocket shoots a second bullet in the backward direction at speed $0.5c$ relative to the rocket. In the earth's frame, is the bullet moving right, moving left, or at rest?

24. The rocket speeds are shown relative to the earth. Is the speed of A relative to B greater than, less than, or equal to $0.8c$?

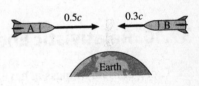

37.9 Relativistic Momentum

25. Particle A has half the mass and twice the speed of particle B. Is p_A less than, greater than, or equal to p_B? Explain.

26. Particle A has one-third the mass of particle B. The two particles have equal momenta. Is u_A less than, greater than, or equal to $3u_B$? Explain.

27. Event A occurs at spacetime coordinates (300 m, 2 μs).

    a. Event B occurs at spacetime coordinates (1200 m, 6 μs). Could A possibly be the cause of B? Explain.

    b. Event C occurs at spacetime coordinates (2400 m, 8 μs). Could A possibly be the cause of C? Explain.

37.10 Relativistic Energy

28. Can a particle of mass m have total energy less than mc^2? Explain.

29. Consider these 4 particles:

Particle	Rest energy	Total energy
1	A	A
2	B	$2B$
3	$2C$	$4C$
4	$3D$	$5D$

Rank in order, from largest to smallest, the particles' speeds u_1 to u_4.

Order:

Explanation: